COWBOY

IN A CORPORATE WORLD

37 Years of Life & Lessons on
Koch Industries' Beaverhead Ranch

To Mardine Brower

Ray Marxer

Ray Marxer

2022

Copyright Information

Cowboy in a Corporate World © Copyright 2022 Ray Marxer

All rights reserved. No part of this publication may be reproduced, distributed, or transmitted in any form or by any means, including photocopying, recording, or other electronic or mechanical methods, without the prior written permission of the publisher; except in the case of brief quotations embodied in critical reviews and certain other noncommercial uses permitted by copyright law.

Although the author and publisher have made every effort to ensure that the information in this book was correct at press time, the author and publisher do not assume and hereby disclaim any liability to any party for any loss, damage, or disruption caused by errors or omissions, whether such errors or omissions result from negligence, accident, or any other cause.

Photography Credits: Images and illustrations are used by permission, and are copyright © Susan Marxer. smarxer@saddlescenes.com unless credited otherwise. Photo credit ©Rene Heil 2007. Personal photo of Sue in Chapter 8.

Bible quotes are courtesy of The Authorized King James 1611 Bible which has no copyright. Verses and other quotes used in this book are public domain, and free of any restrictions.

For more information, email

corporatecowboy@raymarxer.com

ISBN:

979-8-218-01891-7

979-8-218-01892-4

979-8-88759-078-3

979-8-218-06644-4

Dedication

I dedicate this memoir to Sue, my bride of forty-one years. Her shared commitment has resulted in a blessed life, and this book. Also, thank you to my favorite Matadors—our children Clayton, Kristy, and Anna for their help, inspiration
… and moose warnings.

Author's Note

I worked for Koch Industries, Inc. on their Beaverhead Ranch for over half of my life. This book is a memoir written from my perspective, as honestly as memory allows. Others may remember the events and stories differently. In thirty-seven years, many people came and went, and all contributed in one way or another—some more than others. I mention by name key individuals who influenced and contributed significantly to the success of Montana's Matador Cattle Company, likewise many individuals who were instrumental or influential in my own development and success. Not all characters are mentioned by name. Any perceived slight is purely unintentional.

Table of Contents

Introduction --- 1
Prologue --- 5
Chapter 1: Matador Cowboy --- 10
Chapter 2: Getting Grounded --- 19
Chapter 3: Weaning At The Staudaher --- 31
Chapter 4: Sage Creek Foreman --- 40
Chapter 5: Range Restoration --- 50
Chapter 6: Upside Down And Inside Out --- 59
Chapter 7: New Life—New Job --- 71
Chapter 8: Cow Foreman—Toughest Job On The Ranch --- 84
Chapter 9: Cow Camp Summers --- 99
Chapter 10: Computers, Spreadsheets, And Crossbreeding --- 114
Chapter 11: Moving To The Big House --- 127
Chapter 12: Koch Beef Venture --- 142
Chapter 13: Environmental Stewardship --- 158
Chapter 14: Roles, Responsibilities, And Expectations --- 172
Chapter 15: Who Can Stand Before Envy—And HR --- 190
Chapter 16: Essential Partners—Our Cattle --- 202
Chapter 17: Horses—Our Humble Heroes --- 214
Chapter 18: Who We Have In Life --- 231
Acknowledgements --- 244
More --- 246

INTRODUCTION

"Salt is good: but if the salt have lost his saltness, wherewith will ye season it? Have salt in yourselves, and have peace one with another." Mark 9:50

Some of the best advice I ever received came from Bill Caffey, a lanky, rough-edged leader I respected because he was always straight up and to the point. "Don't be a victim," he admonished the entire group during a meeting at Koch's Wichita headquarters several years ago. "You make that choice yourself."

I had no plans to write this particular book. My intention was to write ranch stories, the history, the characters, the land, raising my family in a unique and extraordinary lifestyle—the fun things that made life golden for the bulk of my career. But this core story, of a cowboy embedded in a corporate world, kept intruding. I always figured digging-up-bones was a total waste of time and emotion. What's past is past, but a true memoir requires, at the least, a modicum of context.

I buried the ugly parts ten years ago, but dredging up those memories of the last few years of a successful thirty-seven-year career stank as bad as the day I walked away from them. The hurt and betrayal were right where I'd left them. I'd seen too many good people make the mistake of not letting go, a lesson that had caused me to determine I would never live life in my rearview mirror. Even though I've moved on and successfully recovered my identity as "Ray Marxer, ordinary citizen" rather than the job title I'd been identified with for twenty-one years, the digging is still not easy. But there's a lot of gold under those bones!

My retrospection liberated me, restoring my damaged sense of dignity and accomplishments. I made mistakes but never failed to treat the land, the assets, the animals, and my job as if I were an owner rather than a hireling there for the purpose of a title, paycheck, and benefits. Money or prestige never drove me.

I fully invested myself in my job and my responsibility to leave the land and the business in a healthier and more sustainable position than I found it. Not only did the ranch thrive economically, but our conservation efforts earned numerous environmental stewardship and conservation awards, six at the national level. Employee retention on average dramatically improved from two years to eleven. Mr. Koch even illustrated a Market Based Management presentation he made in Austria, with accomplishments Beaverhead Ranch had achieved using MBM principles. I marveled how God had blessed our efforts, the ranch, and our family. I successfully managed the ranch to a higher level of profitability and environmental excellence—right up to the last day.

The morning of what ended up being the day of my sudden "retirement," I clearly remember being stunned by my own calmness. It could only have been the grace of God restraining my inner cowboy from reacting brashly toward the pompous corporate suit I was facing there in my ranch office. Instead, with a sense of peace and calmness that belied my inner confusion and anger, I decided to walk away. Peacefully. Life is too short to live in conflict and compromise of personal values and beliefs. I didn't realize it in the moment, but the amazing grace and power of Jesus Christ, who saved my soul, was still leading me every step of the way—even out of that situation.

"For what glory is it, if, when ye be buffeted for your faults, ye shall take it patiently? but if, when ye do well and suffer for it, ye take it patiently, this is acceptable with God."

1 Peter 2:20

With a little perspective, I realize now that what went on there at the end, ten years ago, was not that unusual in today's corporate world—just more dramatic. Increasingly, for self-preservation, an army of lawyers and HR departments have taken over employee management. Companies like to project the warm-and-fuzzy facade that "people are the source of our success," when in reality, people have become their greatest liability. Both are true. It's easier to move problems around from department to department and hope they find a better "fit" rather than risk retribution by nipping trouble in the bud. Consequently, those problems never stay long enough in one position to be held accountable for the havoc they create before moving on up the ladder.

I've learned over the years that if I'm not trusting God to hold up my three-legged stool—faith, family, and work—my life becomes as unstable as a two-legged corporate ladder and totally out of balance.

This book is not a disparagement of Koch Industries, Inc. For many, many years, that company was good to us. I would never have had the opportunity of fulfilling my dream of living and working on a big outfit and, even more importantly, working alongside my wife, Sue, and our children as a ranch family had it not been for the company. I still have the highest regard for Charles Koch; likewise, several company people who were great mentors, influencers, and friends to me.

Whether they're planning a career on a corporate-owned ranch or working for a corporation, I hope that young people will read this book and go into that job with a little more wisdom than they will learn in any school. It's in the application of knowledge where education begins. Even corporate managers could glean a little insight from these pages, as could most ranch owners.

Thank you for reading this far, and I hope you enjoy the entire book.

Raymond A. Marxer

PROLOGUE

Farm Boy

"The best preparation for tomorrow is doing your best today." H. Jackson Brown, Jr.

It was 1959, fourteen years after the end of World War II, and the same year as the Yellowstone earthquake. The Russians had launched the first spaceship into outer space, but the significance of that feat escaped me. It was my dream to be a cowboy. Even cowboys occasionally get launched into space, but the law of gravity guarantees their trip back to Earth will be a short one.

At six years old I hoped fervently that Richard, my older brother who milked Frosty, the family milk cow, every morning and every night, would teach me how to milk. I wanted that job. Squirting milk into the cat's mouth, and zapping mice on the rock wall, then packing that bucket of milk up the hill to the house twice a day for Mom to prepare the family's milk and butter supply looked like a fun job.

With Rich more than happy to teach me, I'd bring Frosty into the barn at the assigned time each morning before taking the school bus, and again upon returning home. After giving her a pail of oats to keep her happy, I'd pull up the three-legged wooden stool, set the bucket beneath her bag, push my head into the warm hollow in front of her hind leg, and begin the work of extracting the milk that refilled her bag right on schedule. I grasped both fists firmly around the milk faucets of that good old cow and learned the rhythmic squeeze that sent the hot milk slanting into the bucket.

Unknown to me at the time, that wooden stool would be a foundational influence on the rest of my life in regard to balance. The most stable stool has only three legs. Balancing work, family, and my walk with God would become my life purpose.

I learned quickly that cows do not "give" milk. Just like any natural resource, it takes work to extract it. Nonetheless, I loved my new job. I milked our cow every day till the day I graduated from high school. On that day, Dad told me, "I don't think we need to milk the cow anymore." The full-length cast on my leg due to a knee injury in track made that decision easy, I'm sure, since Dad did not want to do the milking himself.

At fourteen, with my dad and brother, I attended a special school for artificial insemination (AI) of cows. Not sure if I was big enough, my folks let me try going along, knowing I could at least learn the concepts. I had to stand on a stool behind some cows, but learning quickly, I earned my certificate. Since then, I have artificially inseminated many thousands of cows. Even now, in retirement, artificial insemination of cows is a key component of what my wife Sue and I offer in our ranch services business.

Typical of family farms and ranches, all five of us children were expected to pitch in where needed. I was doing much of the swathing, baling, and stacking of hay by age fourteen, and operating tillage equipment. During harvest I drove the full grain truck from the combine in the field, to the elevator in town. I started my own small cattle herd as a 4-H project. My folks took me to the bank, and I had to sit down with a loan officer to apply for a loan to buy my cattle, and provide a well-thought-out plan of how I would pay off the loan.

My mom wasn't too keen on my propensity for rodeo or my drive to learn cowboy skills, but that was Mom. The first time I was challenged to consider the validity of my life-long "cowboy" dream though, I hit an unexpected wall that caused me to pause and seriously consider what my counselors were trying to advise me.

It was my turn, as a junior at Cascade High School in 1970, to sit down with the guidance counselor to discuss my future. I was a good student at most subjects, a good athlete, and an excellent student in agricultural classes and activities, for which I had great interest and passion.

"Well, Ray," the counselor began as he flipped through my file lying open on his desk. "You have an outstanding high school record to this point. We need to talk about your future plans to be sure your senior year is used to your best advantage." Turning slightly, he gestured toward a rack on the wall neatly lined with rows of colorful college brochures and military recruitment posters. "Do you know what you plan to do as an adult?" he asked, swiveling his chair back to face me and folding his hands on top of his desk.

I only had one thing I'd ever dreamed of doing, and I was pretty sure it wasn't what he wanted to hear. I paused briefly to reflect on my answer while making a cursory scan of the rack with the striking military images and representations of large green campuses, stately brick buildings, college teams, research labs, and smiling students. I looked back at the counselor and answered him with unswerving honesty. "I'd like to be a foreman on a large ranch."

"What a waste," he muttered before catching himself. I'll never forget the disgusted look on his face and hearing those words. He quickly recovered, then explained, "I think you've set your goals and expectations too low for the talents I see in you. I feel sure you could contribute successfully in whatever field you choose." He reached to the rack and plucked out a few brochures, including one for Montana's land-grant university and college of agriculture, MSU, and handed them to me with advice to check out different career paths and ways to incorporate ag classes into a degree. I thanked him and left. I did not dismiss his advice but mentally filed it away, knowing he felt his advice was in my best interest.

In college I majored in farm and ranch management, two totally different fields lumped together under agriculture. I found most prerequisite college classes to be unrelated, boring, and not practically applicable to my aspirations. I realized that the practical experience I'd gained through hard work, dedication, 4-H and high school ag classes, coupled with my parents' wise guidance and expectations, had already prepared me beyond many of my entry-level college classes. There were no classes or internships in ground-level cowboy skills necessary for entry into a foreman or assistant manager-type ranch position. I felt like I was wasting money and receiving little value.

When I was ready to look for a job, I had some education and a wealth of practical experience in farming and small-scale animal husbandry, but no ground-level cowboy experience. My next steps were uncertain. In my later years as a ranch manager, I spoke to several different college classes aspiring to learn ranch management. Several of the students were in the very place I found myself at their age and asked many good questions.

I tried to impress upon them that it was important to complete a degree since many businesses or corporations will not even look at a resume without one. But, they needed to be ready to learn what schools can't teach—the hands-on, nuts and bolts practical applications, without which a college degree is worthless. A college graduate can spend their entire ranch career as a hired hand if they never learn common sense, cow sense, horse sense, a love for the land, and leadership skills.

In the world of ranching, that often means being willing to start at or near the bottom. No great manager I know started at the top making big money. Before moving into leadership positions and making the type of day-to-day decisions that determine the success or failure of a business, those managers had previously spent valuable time in the trenches.

"He that is faithful in that which is least is faithful also in much." Luke 16:10

CHAPTER 1

Matador Cowboy

"Go confidently in the direction of your dreams. Live the life you have imagined." Henry David Thoreau

It seemed like maybe my high school counselor might be right. I was getting nowhere fast. In September of 1973, I married a gal from California who I met in college, and then we returned to the farm to work with my parents on my grandfather's original place on the Smith River near Ulm, Montana. We bought a single-wide trailer which we set up on the homeplace and spent our first year there. I was not getting any closer to my dream of being a cowboy, much less a ranch foreman on a large ranch. If I was going to be a hired hand, I might as well be one where I had room to grow, so I began to search for a ranch job in earnest.

After a few calls, I received a lead for the Matador Ranch in southwest Montana out of Dillon. Chances were good that they would have some openings since big outfits typically have a lot of employee turnover.

Montana's Matador Cattle Company was referred to as the "Beaverhead Ranch" by Koch Industries to differentiate it from the original Matador Ranch in Matador, Texas, which they also owned. In the 1970s, the ranch sprawled over

240,000 acres of private land, Bureau of Land Management (BLM) allotments, and state lands in the southwest corner of Montana, not too far from Yellowstone National Park.

Most of the Matador's 5,000 cow-calf pairs and bulls spent summers in the Centennial Valley and high up in the surrounding Centennial Mountains, which border Idaho on the Continental Divide. The ranch always branded the calves before turning them out in the spring since a few strays or cow-calf pairs would often turn up on the other side of the mountain in Idaho. Brands were the only way to determine ownership.

Tom Griggs, a talented, seasoned cowman with a "we're burnin' daylight" job intensity, was the Matador cow foreman. Due to the summer drought, he was figuring on weaning calves off the cows earlier than the usual October or November dates. He needed more cowboys and invited me to interview. I met Tom and his wife, Shirley, at their double-wide trailer at Matador headquarters, where we had a good interview. I was happy to accept when Tom offered me a job on the cowboy crew.

1975 - Cow Foreman Tom Griggs on Roany, Manager, Marion Cross branding, seated on the ground (left) in black hat is Ray Marxer. Sage Creek Foreman.

Tom and Shirley's son, Tracy, was home, so I met him that same evening. Tracy was about my age and worked at the ranch while attending college. He had an intensity and roping skills similar to his dad, a bent for practical jokes, and a propensity for colorful outbursts. We hit it off, and despite our opposite personalities, we ended up being lifetime friends. In cowboy circles, Tracy was often called "Roy Raisin," and it wasn't long before Tracy dubbed me "Ray Boy" after John-Boy Walton, the earnest, responsible big brother on the popular TV series, *The Waltons*. I never did like John-Boy—or the show for that matter.

The cowboy crew averaged about twelve to fourteen cowboys. Most of the cowboys were single due to bunkhouse living facilities, and the cow boss, plus a few married men, made up the rest of the crew. They were always on the lookout to hire at least one married man with a wife willing to be the cook in primitive camps with no electricity and a water pump out front. Otherwise, they'd hire a cook, typically an older guy needing distance between himself and the bottle for a while.

Big outfit rules concerning dogs and horses were more of a challenge than they would otherwise be on a smaller place. On the Matador, a cowboy could bring two geldings of their own, but the company provided ranch horses and expected cowboys to ride them. Riding outside horses for other people was not allowed. Since the ranch ran over one hundred head of horses in the active cavvy, (horses assigned to cowboys), and a second herd, of turned-out horses and young horses, no mares or studs were allowed. Too many geldings ended up getting hurt fighting over mares, and studs also caused many problems.

Dogs were not encouraged since very few were good cowdogs. Dogs were especially not welcome during calving since mama cows tend to associate dogs with coyotes and wolves, their natural predators. Families lived close to each other—mostly in a couple of trailer courts at headquarters, and there was a feedlot across the road. Dogs were to be under control at all times and not kept in houses or riding in the cab of company vehicles.

I understood the mare rule but was not happy about leaving my good mare. Later, at Sage Creek where we didn't have a large cavvy or crew, we had a little more leeway. I had a good dog, but mostly I used him when moving bulls. I didn't want him learning bad habits, especially when dealing with a lot of people and animals living and working in one place. Too many people think that just because they live in the country their dogs are free to roam.

My first day at my new cowboy job was October 8, 1974. Tom and I left headquarters early that morning in the dark. He told me to throw in my saddle, gear, and bedroll because we'd go first to the Centennial, where he and I would ship some cull cows—mostly old cows with severe health or physical defects—out of the Staudaher shipping pens. Then we would take more horses and head over to Sage Creek to help the crew spend the next three days weaning calves off the cows and loading trucks.

When we got to the Staudaher Cow Camp, Tom caught his two horses. They were ranch-raised horses out of Thoroughbred mares and Percheron studs. Up until 1968, that was what the ranch raised. Those horses were big and stout and really broke. Bisco and Blue were the two that were there that day, and he had me throw my saddle on Bisco, who required letting my cinches and bridle out to the very last hole.

I got along fine with Bisco, but being new, I had no idea of the talent and athleticism of that big gray horse. It wasn't until later that I realized he was probably one of the best head horses and all-around stock horses I'd ever had the privilege to ride. I got to ride him quite a bit in later years for roping. At 1,450 pounds, Bisco could cut a cow as good as most 900-pound cutting horses. He was stout enough that you could rope a bull on him, and Bisco would handle him with ease. Tom's younger son Tom placed three years at the State High School Rodeo in the cow cutting event on Bisco. In 1968 the ranch decided to buy horses rather than raise them and sold all the mares.

Once we'd finished our shipping that day, Tom and I headed for Sage Creek, a forty-five-mile drive pulling a trailer, or about twenty-five miles through the backcountry where we trailed cattle or horses. The Sage Creek Ranch was a series of large barns, shops, and buildings laid out like an old western town.

1986 - Sage Creek headquarters *Susan Marxer Photo*

Originally the place was owned by Cooke Sheep Company and had run about 21,000 head of ewes, an enterprise the Matador continued at increasingly smaller numbers until the fall of 1975 when the last two bands totaling about 4,000 ewes were loaded on trucks and shipped out.

Due to the drought, the cow numbers that we would spend the next three days weaning had been reduced to 1,400. The crew laid out their bedrolls on cots in the bunkhouse, which originally was a store and commissary with an old wooden outhouse out back. The cavvy of horses they'd trailed through the hills was five miles up the road by the shipping pens, corralled in a big water-gap trap. We would drive to the shipping pens in the dark each morning to catch and saddle them before starting our day's gather for weaning. The crew would sleep in the bunkhouse, and we'd eat all our meals in the main house with Sage Creek Foreman, Gary Bergner, and his family. The home doubled as a cookhouse through 1990.

On the first day of weaning, my stomach was in knots since this was the cowboy day I'd dreamed of, and I was so afraid I'd screw it up. Sleep was short, breakfast was early, and it was dark-dark. My anxiety about acting like a gunsel had gotten the better of me, and I was one of the first to leave the table, going outside, where I promptly lost my breakfast. I wanted to be a help, not a hindrance, and was excited about being part of a cowboy crew at a big outfit weaning—a whole new experience for me.

We threw our saddles and tack into the back of a pickup before hopping in ourselves. It was still dark when we got to the pens, with barely enough early light to make out the individual horses and cowboys. Gary wrangled the thirty or forty horses out of the water gap to the corral, which wasn't very big, and

Tom started roping horses out for the crew. He roped a little white horse out of the bunch and led him over, handing him off to me. "He ain't real pretty," Tom said, "but he's darn sure a good one. Name's Alpo." Before I could get Alpo bridled, another horse ran into him, and he got away. Mentally I kicked myself for such a gunsel move. The crew kindly helped me get him caught back up again, and I got him saddled and bridled with no more trouble.

In the meantime, while Tom was roping other horses, a big bay horse called Jipp, who was built like a brick, came charging out of the bunch in the dark and ran right over the top of Tracy—just pancaked him. Tracy, aka "Roy Raisin," jumped up, cussing in his typical colorful vernacular. "You auger-headed son of a b—!" he yelled to no one in particular since Jipp had already disappeared back into the bunch. Our day was off and running.

Once everyone was saddled up, we hit a fast trot to the upper end of the five-mile-long pasture, toward the Knox place in the Upper Highbridge pasture. We gathered about 600 pairs toward the pens on the lower end. About a mile above the pens, we had to cross Sage Creek with any cattle still on the west side of the creek. There was no bridge, but generally, that didn't pose much of a problem. On this particular day, several calves wanted to go anywhere but into the water. We had to rope a half dozen or more, drag them across the creek, then heel them to get the ropes off and turn them loose. I felt a little better and hoped I'd redeemed myself a bit by doing my part in getting them roped quickly and efficiently.

We got everything corralled, and weaning was officially underway. Sage Creek pens were large and had one long, wide alley with cross gates that allowed splitting it in two. On the west side was one big pen, and on the east side were numerous smaller pens to sort into. We ran two alleys and did the sorting by horseback, with cowboys on foot running gates. Tom, Ron Weekes, and a couple of other guys sorted cows off, and it went pretty fast. They'd sort cows off into the big pen and calves into a side pen.

Marion Cross, the ranch manager I'd met for the first time just that morning, and the rest of us would sort the heifer calves from steer calves in the other alley. Much of the time, that sort can be just one calf at a time. Marion did it horseback, and it was readily apparent his horses were well broke. Those two big black horses could really cut a cow.

1986 - Marion Cross sorting calves at Staudaher Pens in the Centennial.
Susan Marxer Photo

The first sort gate usually carries the most pressure because the animal needs to be stopped at the first gate if there is a mix-up. Marion asked me to run the gate for him, which was the start of a special relationship. We both gained respect for each other that day. I was just a twenty-one-year-old green kid, but we didn't have one mix-up. I ran the gate for him for a couple of hours that morning, stopped for lunch, then we went back to sorting in the afternoon.

Marion sorted for a while, then said, "Ray, why don't you get your horse and sort the next alleys." The very idea that he had confidence I could do the job spoke volumes to me. I got my horse, but felt a little uncomfortable because there were other cowboys there that were better hands than I was and certainly older and more experienced. But he was the boss. The other thing inspiring to me was that he didn't just get off and watch; he ran the gate for me. Marion and I teamed up to work cattle every chance we got from that day on. Marion was the manager for fifteen years while I was there and a great mentor to me.

Once we'd finished the three days of weaning at Sage Creek, a few cowboys, including me, trailed the horses the twenty-five miles cross-country to the Staudaher Cow Camp. Leaving Sage Creek, we headed out through Big Basin, Little Basin, and Clover Creek to the Blacktail, then to the Staudaher Cow Camp in the Centennial Valley, where we would wean the main herd later in the month.

There's something about hitting a long trot while pushing a herd of horses out through rugged, primitive, sagebrush hills, dodging badger holes, bogholes, and treacherous hidden barbed wire from long since abandoned homestead fences, that makes the soul sing. I was right where I wanted to be, in the element of all I'd dreamed of and learning everything I could soak up.

I'd passed my first big test and was officially a "Matador Cowboy."

CHAPTER 2

Getting Grounded

"Attitude is a little thing that makes a big difference."
Winston S. Churchill

1975 - Tracy Griggs dragging calf, Ray Marxer on right.

Cow Boss Tom Griggs didn't talk much, and when he said something, he expected his crew to listen up. I couldn't have picked a better leader to learn from. He led by example, and anyone could absorb a lot just by following and showing interest—he was happy to teach anyone who had a genuine drive to learn. Work began before sunup, and you knew it would be a long day if he stopped at the commissary and came out with cans of Spam or corned beef and a box of saltine crackers, or he might bring Vienna sausages, peanut butter, and a loaf of bread. He drove with his window down, no matter how cold it was.

Cowboys did most of their work from the back of a horse, covering countless miles throughout the year. The only horse trailer on the whole ranch had been there less than a year. It was a big gray monster, thirty feet long, five and a half feet wide, and six feet tall on three axles. Loading that trailer was like loading a horse into a culvert since most of the ranch's horses were part draft and big and solid. We got the chance to modify the trailer a few years later when the cook at the Jake Cow Camp got desperate for a drink and headed to town with the ranch truck still hitched to that big gray monster. He took out part of a gate when turning onto the road, and during repairs, they did a little remodeling, cutting ten feet off the trailer and taking out one axle.

When I hired on that fall, Tom Griggs, the cow foreman, had a single-cab ranch pickup, which meant the cab was packed to bursting with bodies. Anyone that didn't fit would ride in the open truck box. Once they got that big gray ranch trailer, cowboys could ride in the nose of the trailer. It was bitter cold in the winter and hot with choking dust in the summer. We'd use any saddle blankets lying around in the nose to mitigate the discomfort, even putting them over our heads to filter some of the stifling dust. After the rig rattled to a stop, horses would need to be unloaded before the cowboys in the nose of the trailer could exit.

In 1976, the last year Tom would be cow boss, he was finally able to get a crew-cab pickup with an entire back seat and two more doors. It was still a race to mash as many bodies into that truck as it would hold and still be able to shut the doors. It was too miserable for anyone who ended up in the truck box or trailer nose to get in much sleep during the hour-long drive to and from the Centennial.

Even so, the upgrade in transportation was a big step up from the years before 1974 when I came on the scene. An old red snub-nosed Ford stock truck with a rack on the back was the primary mode of horse transport from headquarters for several years. The ranch mechanic had fabricated a special tailgate for the rack that extended a couple of feet higher so a saddled horse could be jumped in and out.

Before they got their first horse trailer in 1973, the only way to get anywhere without having to ride or trail the cavvy was to haul them in that straight truck. They just backed up to a dirt or rock bank for loading and unloading to jump the horses in or out. A few makeshift loading docks were scattered around the ranch, but usually, a bank was handier. Most of the time, they trailed the horses. Fitting a crew into that red truck meant a couple of cowboys would have to load up in the back—in the stock rack with the horses. Aside from being very uncomfortable, riding in a stock rack with animals can be pretty dangerous.

The cowboy crew's second vehicle was for the foreman at Sage Creek. Sage Creek Ranch was a stand-alone camp separated from the main headquarters by a mountain ridge. Usually, someone at headquarters handed down an older pickup to Sage Creek when they upgraded. Sage Creek had no horse trailer or any other way to haul animals. In December 1974, the company got a sixteen-foot neck-over, canvas-top trailer intended for Sage Creek, but it spent most of the first couple of years at headquarters because it was so handy.

A third ranch vehicle was for "Smitty" Harold Smith and his wife, Naida, who cooked for the crew at the Staudaher. Smitty was the jigger boss under Cow Boss Tom Griggs. The truck was a one-ton dually Ford two-wheel drive with an Omaha stock-rack bed. Like the red stock truck, the driver backed up to a bank to jump a horse in or out.

After the fall work settled down that first year, I was able to spend a lot of time with Tom and a few of the other crew, sharpening our cowboy and horsemanship skills. Tom was an avid team roper. That winter, he would offer to take any of us interested cowboys to little jackpot team ropings around the local area that he was going to. Tom would mount us on his top-notch palomino horses Nip and Tuck, who could do any event—calf rope, head or heel in team roping, or even bulldog. The capabilities of Tom's horses were an indication of what a horse hand he was. I appreciated and enjoyed Tom's help and mentoring in calf roping. He complimented me once by saying that my calf roping reminded him of Warren Wuthier, a highly-skilled friend from Wyoming.

On occasional days off, Tom would haul his horses and a few interested cowboys over Lemhi Pass to Leadore, Idaho, to team rope during the winter in one of the few indoor roping barns in the area. The pass could be pretty treacherous in the winter. Sometimes we would need to unload the horses and ride to the top of the divide so the truck and trailer could make the hill. The big roping at the Winter Fair in Bozeman was another team roping we went to that first winter. I teamed up with Matador cowboy Phil McChesney, and we placed well.

Roping—both team roping and tie-down calf roping— were skills I developed that benefited my ranch duties and provided an avenue to enjoy life away from my job. Tom Griggs helped me immensely in developing those skills. The roping activities led to our involvement in building a roping arena at Dell and also participating in, and helping with, the Labor Day Fair and Rodeo in Dillon. We put on gymkhanas for kids and taught several kids to rope.

I was hungry to learn not only cowboy and livestock skills, but the whole story and history of the Matador Cattle Company. I grew up in Charlie Russell country—as in C.M. Russell, the famous cowboy artist. *Square Butte*, a hallmark of many of his paintings, was a landmark in the Great Falls area. It was exciting to me how young the wild history of Montana was, and I loved listening to stories from the local old-timers.

Between Tom and Marion Cross—mostly Marion—I learned a lot of fascinating ranch history. In the early years, when Marion's brothers, Jim and Jerry Cross, were cow foremen in the 1950s and 1960s, the only vehicle they had was an old Jeep pickup. They had a tin-sided cookshack that they pulled around to wherever they were working cattle. The cowboys trailed their horses and camped out with them. The old cookshack sat at Sage Creek headquarters for a while after it was retired, then we tore it apart and discarded it. Later I used the axles and some other parts to build another trailer.

The Matador ran as many as 12,000 sheep before dwindling to the 4,000 ewes to be sold the fall of 1975. Sheep had been a significant enterprise for a couple of decades, requiring many sheepherders, sheep-wagons, dogs, camp tenders, and a full-time sheep foreman who was assigned one of the few ranch vehicles. When Marion Cross was the sheep foreman in the 1960s, he had a Jeep Gladiator pickup with a stock-rack on the back and a well-broke horse. He could haul his horse just about anywhere in the back of that pickup.

Marion's camp tender, the guy responsible for moving the different sheepherder wagons or their camps to the next base, also had a ranch vehicle. The camp tender was an essential liaison for the remote and nomadic sheepherders. He visited or moved the camps weekly, brought groceries and necessities, mail, dog food, and anything else that might be needed or ordered. He checked that the herders were doing all right and that their horses were sound and had shoes. By 1974 there were only two bands of sheep left.

One of the bands had a little Mexican fellow named Pete, who was their herder. I believe they spent the summer either at Mount Carey allotment in the head of the Ruby by the Divide Creek Ranger Station or at Hell Roaring in the upper end of the Centennial Valley. An older gentleman named John herded the second band of sheep. He and his band summered at the Ben Holt place, up on top of Blacktail Ridge at the head of Jake Canyon. The camp was very remote and high, at about 8,500 to 9,000 feet.

I was never involved with the sheep at Sage Creek until June of 1975, when they sheared the sheep at the old shearing plant and sheep corrals that were on Basin Creek. The plant was about six miles straight east of the Sage Creek headquarters at the very bottom of Big Basin. There was a gigantic old shearing plant that had initially been a log homestead-era cookhouse and bunkhouse, and there was a big wool warehouse. The warehouse was just a rusty tin building but quite large. It tied into the center of a very long, low roofed structure where the shearing of the sheep occurred.

A steel shaft, about an inch and a quarter thick, ran the entire length of the long shed up under the roof. A little one-lung gas engine drove the shaft. Part of the far end of that shed was gone, but there were still twenty-seven separate shearing heads that came off that shaft. Twenty-seven men could work at the same time shearing a band of sheep. There were wood platforms under the eight-foot-square stalls where each shearer worked, so they could keep the wool clean.

Four wool stomping stations were between those shearing stalls in the open end of the warehouse. They consisted of a wood stand about nine or ten feet high—a little higher than the woolsack was long. It had a wood deck on its top, with a big recessed circle cut into it and a metal ring with hooks. An empty woolsack was dropped through the hole, and the top—the open end of the woolsack—was folded over the metal ring and hooked to hold the woolsack in place. A person would get into each of those sacks and stomp the wool down with their feet. Once they had three fleeces, the stomper would begin stomping three at a time in the sack. He would have to push the wool fleece past himself in that sack and keep stomping. The goal was to try and get as much wool in a sack as it would hold. One woolsack would hold around 320 pounds when stomped in by foot.

In later years, as we did in 1975, a hydraulic sacker was used. I could get up to 400 pounds in those woolsacks, but they were extremely tight. When the sacks were full, a pole would be used as a lever to jack up the weight off of the sack opening so it could be unhooked from the ring and tied off. Someone would take a needle and sew the top of the sack, leaving two "ears" on the top corners so a person could get a hold of it. Then, a couple of crew members rolled the sack out of the way into the warehouse, and someone else would start stacking full woolsacks in the warehouse.

The ranch operated the sheep-shearing shed using that system for years until the last sheep were gone. The shearing shed was probably installed there around 1910. Old-timer Dan Peterson, a neighbor and friend who had grown up around Sage Creek, spent much of his life around that shearing shed. He told me that one of the reasons the country looked so tough in Big Basin and why the shearing shed was so big, was because they had to shear an entire band in one day and move it out of the way. Another band of sheep would already be waiting to be shorn the following day. Dan told me that typically, the shearing shed would run about one month per year—most of June. They tried to shear sheep every day, but rainy days could delay the shearing.

In those days, the bands of sheep numbered between 1,500 and 2,000 ewes which would equate to 45,000 to 60,000 sheep sheared there every year—a lot of backbreaking work. The year I helped in 1975, we sheared sheep there, but the ranch had a portable shearing unit come in. Some of the guys were from Lewistown, Montana, and five or six guys were shearing simultaneously. They sheared all 4,000 head in two or three days.

We sorted the lambs from the ewes so the ewes could be shorn. I had never been around sheep sorting. The sheep pens were set up with smooth, wood-sided sorting alleys, three and a half feet high and narrow enough that a sheep couldn't turn around. The alleys were long and had concrete bottoms to prevent the deep and muddy ruts like the old-timers had when working on dirt. The alleys had "dodge" gates to send ewes into one pen and lambs into another. A lot of sheep people still use the same type of sorting alleys.

Marion Cross was there, and I got to watch someone who was an expert at sorting. Marion ran two dodge gates. The alley went out straight, so he could sort three ways while standing in one spot. As the sheep came up the alley, Marion could sort ewes to one side, lambs to the other, and if there was a stray or a sheep that needed special attention, he could let it go straight out the end of the alley. At the same time, he had a clicker counter in each of his hands for the two different gates, so when those sheep came through, he could sort and count as fast as the sheep moved through. It didn't take long to sort them, and when Marion got done, he knew how many were in each pen which was pretty amazing to me.

The shearers began shearing the ewes as soon as we got them sorted. While the shearers were busy shearing and packing wool bags, Marion and I ran the lambs into the scale, packing them in as tight as possible. We went into the enclosed scale with cans of paint and branding irons, and we put the ranch's "wagon-rod" sheep brand on each lamb. We used different colored paint for each band's lambs, and after the shearers had finished the ewes, we repeated the process. There were still many sheep in the country, and it was necessary to paint brand the ranch's sheep before they trailed to summer country. The sheepherders could sort the sheep according to paint brands if the bands got mixed with somebody else's sheep.

All 4,000 ewes and their lambs would be sold early in the fall. Both bands returned to the same area as the shearing shed. There we would wean the lambs off the ewes, weigh them, and load them on semis. Gale Storer from Idaho was the trucker, and he could load about 500 lambs on each truck. He had a little rat terrier dog that was fun to watch as he worked the sheep. Just the two of them, Gale and his dog, could load 500 lambs on a semi.

As soon as the lambs were loaded, we loaded up the ewes. On the last truck, Marion made an agreement with the sheep buyer to load all the sheepdogs, except two favorite dogs that the herders wanted to keep, into the back compartment. The dogs would go to a new home and a new job.

One of the unfortunate accidents that week occurred because the camp tender, which the ranch hired for the summer, slacked on his job. He knew the sheep were going away, so he didn't have much motivation to do a good job of camp tending. The herders got by okay, but the horses got shortchanged. The camp tender was responsible for keeping the sheepherder's horses shod so they didn't get sore-footed, but we found out later he'd failed that job miserably.

John, the older herder who'd stayed alone with his sheep all summer, high up in the Ben Holt, had done an excellent job, bringing in fat and healthy choice lambs. Both his and Pete's bands were settled in Big Basin the day before we shipped the sheep out. John's job in the high, rugged country was complete, and in one more day, he would be done for the summer. He was down off the mountain, tending his sheep for the last day—when his paint horse fell with him and broke John's leg.

We got John to the hospital and took care of everything, but I felt terrible for him. I went up the mountain to get John's camp and horses. The camp tender, who'd already quit, had done such a lousy job of tending the horses that when I picked up the horse's foot to look at the shoes, I could see that they had been on so long that the horse had worn the center of the toe of the shoe entirely through. There were just two little strips where the nails were on each side of the front feet. The shoes had probably been on those horses all summer. I'm sure that's why the horse fell.

The significance of that day didn't sink in until later. I had just witnessed the end of a historical ranch's sheep era. Most of the sheep operations in southwest Montana and across the country were also cutting back or switching to cattle. The demand for lamb meat for human consumption had dropped drastically. The increasing use of synthetic textiles like polyester and nylon for clothing rather than natural wool changed the market outlook for products that sheep provided.

In the 1920s, Montana's Beaverhead County alone had over 400,000 sheep. In 1918, The Cooke Sheep Company, which built the headquarters at Sage Creek, and were the earliest owners before Koch purchased the ranch, had 21,000 ewes. They grazed sheep in the Big Hole, at Sage Creek, and clear to West Yellowstone in the summer.

Ranches that ran more sheep than cows did so partly because a lot of the high country is better suited to sheep than cows. It's too bad we've lost the sheep component because their grazing habits are different than cows. Managed correctly, a combination of sheep and cattle grazing helps keep a balance between forbs, grass, and brush—a balance that contributes to a diversified, sustainable, and healthy range. Sheep also kept wildfire fuels in check and attracted sage grouse—probably because the sheepherders controlled predators.

I never had any desire to work with sheep, but Marion Cross was good at it. He was one of the most patient men I've ever known, and he never said much when working with a crew, nor did I ever hear him raise his voice. Even when he went to his pickup to get his lunch one cold day of weaning—and discovered that one of the crazier cowboys had already beat him to it, leaving nothing but the empty wrappers.

These were all things I would learn over the next few years. For the fall of 1974, I was still a new Matador cowboy about to experience my first weaning in the Centennial Valley.

1989 Weaning at Sage Creek pens. Standing outside on left: George Munski, Steve Ingram, Marion Cross, Clayton & Kristy. Crew includes Ray Marxer, Jim Bishop, Pete Slater, Mike Cox, Bruce Sandifer, Bill Whittenburg, and others. *Susan Marxer photo*

CHAPTER 3
Weaning at the Staudaher

"It was a bit of the Old West that was still alive, where cowboys were still wild and woolly, and horses were tough."
Ray Marxer

In the summer and fall of the year, most of the cowboy work isn't near ranch headquarters. Cattle begin moving to the high summer country as early in the spring as possible, with the goal of having everything placed by the Fourth of July. We'd brand and trail in bunches that were designated for the different pastures, primarily in the Centennial Valley, Sage Creek, or the Upper Blacktail. Bulls were turned in with the cows by then, and most everything was trailed, averaging ten to twelve miles per day up the Blacktail Road. Cattle destined for Sage Creek on the other side of the mountain were started up the Blacktail then trailed west up Red Canyon.

1986 - Cowboy Crew moving to Staudaher Cow Camp - Eric Soito, Rod Richardson, Don Reese, Craig Filmore, Ray Marxer, Dean Deide, Clayton Marxer
—Susan Marxer Photo

Most of the crew that had been there through the summer had lived in one of the camps, which included the main Staudaher Cow Camp, the 7L Camp on the south end of the Centennial, and Sage Creek, on the west side of the Blacktail Mountains outside of Dell, Montana. Cowboys from the Centennial could return to headquarters on the weekend. Since cow camp facilities for families were limited, some married cowboys were expected to spend the summer at headquarters running hay equipment.

There was no electricity or even a well at the Staudaher. The water came from two little creeks that had their source up in Peterson Basin and ran through the camp. An old, weathered shack served as the cookhouse. It had a little walk-in porch with a tiny pantry, a wood box, and a propane refrigerator. Inside was a homestead-sized kitchen and dining area, and off to the side, was a little room for the cook's quarters. It was one of those old tar-paper-sided shacks with a tan-colored brick print. During winter, the varmints loved it, and some tried to maintain residence during the summer as well. Skunks and mice were perennial pests.

Smitty and Naida had been living in the cookshack all summer, and Smitty kept his shotgun handy since skunks were habitually scratching around the shack looking for mice or the food they could smell. One morning, Ron and I were the first to show up for breakfast and were greeted by Smitty standing casually by the table with his shotgun cradled in his arm. "Well lads," he drawled, "we have us a little chore before breakfast."

We could smell the hot, crispy potatoes frying in a giant, cast-iron skillet on the propane stove about four feet away, and were of a mind to fill our plates. The half grin and the shotgun did not inspire confidence in what Smitty had in mind, but we were on his turf. Skeptically, Ron took the bait. "Whatcha need?"

Smitty pointed to the propane refrigerator that sat in the little pantry right off to the side of the door. "If you lads will just tip that back, I'll shoot mister skunk." Ron and I shot each other a questioning look. Smitty's "little chore" could not possibly end well. Propane fridges are heavy, and besides, there wasn't really a good way to get a shot off inside the shack. Exposing the skunk would trigger his self-defense, and shot or not, he'd spray inside the cookshack. We'd much rather leave him alone until after breakfast.

Undeterred, Smitty was already positioning himself in the door where he didn't have a lot of room, but he had a straight shot. Reluctantly, Ron and I obliged Smitty. We got a hold of the top of the fridge, slid it out just far enough to tip it back so the skunk would run out, and BOOM! Mister skunk's scavenging days were over.

Smitty grabbed the dead skunk by a leg and dragged him outside. The shotgun had guaranteed full release of mister skunk's thick choking spray. Our eyes were watering, and we were all gagging as the rest of the crew came choking into the shack to eat breakfast. There was no talking or lingering that morning. The penetrating smell faded a bit but never went away the rest of the fall. We just got used to it.

Back up the creek, between the cookshack and the barn, were four or five little shacks that were probably former railroad camp shacks that had been hauled in for housing. They measured about ten by twenty feet and had been constructed with two-by-four studs covered by ship lap lumber on the exterior. The roof was domed. Inside the door was a rough wood floor, a small window in the back, and just to the left of the door was a little potbellied stove for heat. There was room enough for three cots in each.

That first fall, Tracy Griggs and I stayed in one of the shacks. During weaning at the Staudaher pens, Gary, the Sage Creek foreman, and his support help, Bruce Corman, came over to help, staying with us in camp for that week. Bruce, who was about my age, had been on the ranch a long time. His dad was Big John Corman, who had been ranch manager for several years in the 1950s through the early 1960s.

Bruce moved into the shack with Tracy and me as we were readying our gear for the start of weaning the next day. The lunch bell rang, and as soon as he ate, Bruce made a beeline back to our shack and his bunk. Tracy quickly cleaned his plate and excused himself, saying, "I think I'll crash too. It's too danged cold to do anything else."

I was only a couple of minutes behind him. As I approached our shack, I heard an explosive commotion, and here comes Tracy, bounding out the door like a fox out of a chicken coop. Right behind him, Bruce burst through a fog of smoke, hot on

Tracy's heels. I ran to the shack expecting to find a fire, but the only fire was in the stove. A pall of dark smoke hung in the air, along with a suspect fuel smell.

I don't know if Bruce caught up with Tracy, but when Bruce came back, I noticed his eyebrows were tipped with tiny, tight yellow curls. Tracy came in right behind him, snickering with his "Roy Raisin" pleased-as-punch observation. "About time someone lit a fire under you!"

Still under his adrenaline rush, Bruce let fly a string of expletives directed at Tracy. "You sorry snake-bit son of a b—!"

"What happened?" I asked, hoping to defuse the tension before fists started flying in that tight shack.

Tracy was more than happy to fill me in. "It was like a freezer in here, so I was building us a little fire." He pointed to a spray can lying outside the open door with a grin. "I figured a little starter would light it right up."

Never one to miss a good opportunity for horseplay, Tracy had picked up a can of ether sitting by the door that was used to boost hard-starting engines. He sprayed a bunch of that volatile mist into the cold stove and all over the laid-out wood, then tossed a match into the open stove door as he stepped out of the shack. A big ball of red flame exploded from the stove as the spray ignited. Bruce wasted no time exiting through the fire to get outside. The flash of flame disappeared as quickly as it ignited, but not before Bruce singed his eyebrows.

1989 Staudaher Cow Camp crew moving cattle around for weaning. 2nd from left is Pete Slater. Jim Bishop behind in slicker, Colt Haugen, Bill Whittenburg, Bruce Sandifer
Susan Marxer photo

It was a bit of a throwback to the Old West when cowboys were still wild and woolly, and horses were tough, receiving most of their training on the job. As long as the food was decent, a cowboy pretty much took everything in stride. When he got tired of the job or crossways with the crew or boss, he simply picked up his gear and bedroll and moved on.

When Charles Koch was a teenager in the 1950s, he and his brothers spent some of their summers working at the ranch. It was in one of those shacks where he was bunking with a mountain-man type character called Bitterroot Bob. Years later, Charles would tell how Bitterroot Bob scared him half to death by shooting off his loud pistol inside that shack with Charles only a few feet away. It undoubtedly set his ears to ringing and left a lasting impression and a hole in the roof.

Charles would enjoy repeating that story while reminiscing with ranch cowboys about his younger days on the ranch. Both Charles and David Koch enjoyed swapping stories with the crews when they'd visit the ranch—which usually included a potluck-style cookhouse meal at ranch headquarters with all the employees and their families. The old shack Charles stayed in was accidentally included in a camp cleanup years later and was burned down.

The first time I helped wean at the Staudaher, in 1974, we weaned for six or seven days at the Staudaher pens two miles from camp and shipped the weaned calves right off the cows. They all had to be brand inspected to ensure no stray cattle were loading on the trucks with ours. The sheriff and deputy sheriff were also part-time brand inspectors, and longtime sheriff, Lloyd Thomas, had managed the Matador from 1951 to 1952, the first year of Koch ownership. He had a fond attachment to the ranch from the time he'd spent there and decided he would come up and be the brand inspector. He would stay in camp with us for five or six days and inspect brands every day.

Ron Weekes and I cleaned up a sheepherder wagon so he'd have a decent place to stay. It had a small bunk in the back, and by the front door, there was a little wood stove. Sheepherder wagons were amazingly compact and complete living quarters on wheels—the precursor to modern-day campers.

Lloyd was an entertaining storyteller. He had an unending repertoire of stories and an excellent command of the area's history. The first evening it was still light outside, and he was telling stories to some of us young cowboys. Tom was there too. Lloyd was on a pretty good roll, probably helped along by the bottle he was nipping on.

"Come on over to my camp," he suggested since it was getting dark. So, we headed over to his wagon. Three or four of us were sitting in the back of the wagon, and Lloyd was still telling stories. We'd been in there for a while when he decided he'd better start a fire in his stove so he'd have heat for the night. Lloyd stood in the open doorway and began filling that little stove with small chunks of chopped-up wood, never missing a beat with his story. I'm not sure if he realized how full he'd stuffed that stove, but it really wasn't the wood that was the problem.

We were all sitting there listening to him and watching him. But when Lloyd reached around and grabbed the can of lantern fuel we used for Coleman lanterns, tipped up that metal can, and started pouring it on all that wood he'd just stuffed into that tight little space, we began to get a little nervous. He was still talking but stopped his story long enough to caution us, "You fellas might want to sit back." We were all sitting as far back as we could, on the wrong side of the door, when he casually lit the match and flipped it into the stove. Fortunately, the stove held. But I'll never forget the lids jumping six inches straight off that stove when the fire lit. It's a wonder we didn't all get blown up, but Lloyd the sheriff, just went on with his story.

That day was the kickoff to my first weaning in the Centennial Valley and month two of my cowboy experience for a big outfit. One cowboy would leave in the dark in the morning to wrangle the horses out of the pasture into the pen behind the barn. We'd eat breakfast by Coleman lantern light, then fire up the Coleman lanterns in the old weatherworn barn so we could see to saddle our horses.

Even though it was the 1970s, not much had changed in the far-flung camps. Housing, pens, working facilities, cowboys, and horses were tough remnants of the rough-and-tumble era that settled the hard, high plains country.

The headwaters of the Missouri River are formed high in the Centennial Mountains providing good water and summer grazing, but the country is too high, wintry, and arid for farming. Sheep numbers had diminished. In part because it was so hard to hire herders and partly because lamb meat was no longer as popular for the dinner table as beef. Most homesteaders had starved out during the Depression and walked away with what they could carry. Some of the homesteads were bought out by the larger ranches. The Matador's cow camp housing, cookhouses, and barns were mainly original homestead buildings with no electricity or indoor plumbing.

It was an exciting time for young men with cowboy dreams. Bunking in a cow camp or riding out across that rugged, lonesome country on horses that were as tough as the land they traveled, were opportunities few people would ever experience. For many, the romance quickly gave way to reality and contributed to the constant employee turnover. We lived through the end of an era forty years after most populated areas had long since transitioned. The purchase of that big, gray monster horse trailer was a needed step forward, but we still had a long way to go.

I wasn't sure how long we would stay with the Matador. I had no inclination to be one of the married cowboys who would be required to spend summers putting up hay at headquarters. But I never had to make that decision. About four months after becoming a Matador cowboy, I had a completely different decision to face.

CHAPTER 4
Sage Creek Foreman

"A dream doesn't become reality through magic; it takes sweat, determination and hard work." Colin Powell

The calves we'd weaned had been loaded and shipped out on trucks headed to feed yards. Some went to the feedlot at headquarters for backgrounding or replacement heifers. We turned the cows back out to graze until winter set in and the snow got too heavy to forage. By then, they were happy to head home. Once we headed the Centennial cows back down the Blacktail Road, they trailed on their own. The younger cows were started toward headquarters first, followed by the older cows a week or so later. Usually, it was late November or early December.

Horses that remained at the Staudaher Cow Camp for the cleanup work were also headed toward their winter home at headquarters and turned loose once we'd gotten them past the cattle guard at the Jack Thomas place a couple of miles out of camp. Horses would hit a long trot or lope, and could make the thirty-five-mile trip in three hours. Staudaher cowboys moved home to headquarters about mid-November once we had turned the cows back out to graze.

By mid-December, we'd finished getting all the stock home and were wrapping up pregnancy checking, which we did ourselves. Replacement heifers were selected and placed in the feedlot, or maybe on stubble for a while, and we turned pregnant cows out in specific pastures for the winter and for calving. First and second-calf heifers would be fed hay and calved at headquarters. Part of the first-calf heifers, about 500 or 600, were wintered and calved at the Sage Creek headquarters, where the foreman fed them hay on the lower meadows.

All the older cows winter-grazed in the Upper Blacktail meadows, which we reserved for winter use. About 1,000 of those older cows wintered and calved on their own in the Wire Field, a wide-open 14,000-acre pasture with limited protection. Every other day all the older cows were supplemented with fifty-pound protein blocks at the rate of two pounds per day per cow throughout the winter and spring.

Semitrucks delivered loads of protein blocks. It took a regular "Block Brigade" to unload the blocks by hand and stack them in Quonsets. Both the cowboy crew and the farm crew were required to help unload, carry, and stack each block. The cow boss and cowboy crew hand-loaded and fed the blocks out of the back of pickups in the different pastures.

Riding the feedlot pens, moving weaned calves around on stubble pastures, loading and feeding blocks, and other tasks kept full-time crew members busy. Cowboys back then were expected to shoe their own string of about four or five horses. They'd pull the shoes off all but one or two horses during the winter and turn them out barefoot on the Lower Blacktail bench for a winter vacation.

In December of 1974, during the dead of winter and the middle of the feeding season, Marion and Tom approached me about going to Sage Creek to help Gary Bergner. Things were falling apart in Gary's life. He was going through a divorce, his family was gone, and Bruce, his help, had quit. Gary was there alone and doing a lot of drinking. Somebody needed to go up and help him feed, so I spent the next ten days making the 120-mile round trip from headquarters to Sage Creek to help.

Hay was being delivered to Sage Creek by the semi-load from Idaho in the winter of 1974 to 1975. The hay was in small square bales that weighed about 110 pounds each. We loaded sixty-five bales on each of the two trailers by hand, seven days a week that winter, then fed them off by hand. Doing the math, we each loaded 14,300 pounds, or about 3.5 tons, per day, per man—every single day. If the hired help left, one man was left to handle the entire job himself, which I had to do more often than I like to remember.

Whenever a semi-load of hay was delivered that winter, we had to unload the semi by hand and build the haystacks. In later years, Harvey Reynolds from Dillon hauled our hay. He'd bring a tractor and hay loader and leave it there when he didn't need it elsewhere until he'd gotten all our hay hauled into Sage Creek. Harvey let us use the tractor-loader for our daily wagon loads when it was there. If his tractor was elsewhere when he came with a load, Harvey would throw the bales off the semi while we built the stacks. He was a tall, wiry, long-geared guy who could stand in one place and pitch a whole row of bales faster than we could keep up with him. We always enjoyed Harvey with his big, friendly grin and pleasant conversation.

Tom talked to me when I returned from feeding one day, telling me Gary wanted out of there, so they were moving him out. Tom and Marion Cross wanted me to take on the Sage Creek foreman job.

I hadn't expected any kind of advancement that fast. The four months I'd spent working on the cowboy crew at ranch headquarters had been critical for me to get a feel for the ranch and develop skills essential for any cowboy to become a top hand. But taking on the responsibility of the 80,000-acre Sage Creek division was a big step I didn't yet feel qualified for. I was reluctant to accept.

Marion had spent nine years at Sage Creek before taking on the management job at headquarters and he fully understood what was required. At Marion's urging, and because of his quiet confidence in me, I accepted. I would spend the next eleven years gaining experience, skills, resourcefulness, and knowledge to equip me for a much bigger agricultural career than I had ever anticipated.

Bruce Corman returned and helped me for a month or two before he left again. I was able to hire Chuck Wiggins, an older fellow whose dad, John, had been sheep foreman at one time.

He helped us get through feeding and our tough calving that spring.

Sage Creek's harsh winters put our spring work about two weeks behind everyone else. Consequently, other ranchers had already hired most available help before we'd even started hiring. At twenty-one years old, learning to hire and manage support help was a challenge since some employees were three times my age and much more experienced in ranch work. It wasn't easy to find single guys willing to work in such a remote location.

Besides the main house, the only living facility was the old homestead-era commissary building, which we used for a bunkhouse. There was a weathered outhouse behind the bunkhouse. The foreman's wife provided meals in the main house with the foreman and family. An outside door facing the bunkhouse, allowed access to a washing machine, sink, and shower in the house basement for the use of the hired men. Very little in the entire Sage Creek complex had been updated since the 1930s. The house was never locked.

The main house was a big old boxy two-story frame house, with weathered wood siding so loose the wind came right through. Upstairs there were four bedrooms. On the main floor was the kitchen with a small porch entryway and a small bathroom/laundry room by the basement stairs. There was a six-foot by thirty-inch table where the family and one or two hired men had regular meals. The other half of the main floor was a huge, drafty living room which was also utilized as a cookhouse for occasions when the entire crew was there to help.

1975 Sage Creek HQ - house and bunkhouse with old loose siding. Ray Marxer, Sage Creek Foreman

The house had served as a cookhouse during the sheep-ranching era. Marion and his wife, Jackie, lived there with their three little girls for nine years, and Jackie would feed up to thirty herders and shearing crews during lambing and shearing seasons, in their living room. The hands would come in at different hours, so she constantly cooked and cleaned.

Marion and Jackie Cross circa 1990

Dell, population twenty-eight, was ten miles down a rough gravel road with access to I-15. It was another forty miles north to Dillon then ten more miles east up the Blacktail to get to headquarters. Several employees that drifted in and out for work at Sage Creek didn't own a vehicle. The bus line between Salt Lake City, Utah, and Butte, Montana, would stop at the Dell Mercantile, which also served as the post office.

My wife at the time was a good hand, which made it easier to get things done even when we were between cowboys. I'd learned a lot from her dad, who managed the Rancho San Juan Ranch in the Los Alamos area of California. During working visits, I'd also had the opportunity to work alongside great hands, such as the Branquinhos, who would become lifelong friends. Henry Millard, an older gentleman with a very slight build, was a great horseman. He stirred a desire in me to develop a level of communication with horses, as he quietly demonstrated. My boss, Tom Griggs, was a great cowman and horseman. The influence these folks had on me for developing horsemanship, cattle working, and roping skills, was of great value to me for the rest of my career and to this day.

That first spring of 1975, we had about 500 first-calf heifers to calve at Sage Creek. The cowboy crew would calve approximately 900 head at ranch headquarters in the Lower Blacktail south of Dillon. At 6,525 feet elevation, Sage Creek averaged fourteen degrees colder than headquarters, which sits at 5,500 feet. The wind was more constant as well. While the wind was unpleasant for us, I would learn that it was a benefit for the sheep ranchers in the past. The wind blew the ridges bare of snow on the expansive range so they could winter bands of sheep with less supplemental feed.

The calving facility they'd used for a few years consisted of a couple of smaller wire traps, about 300 feet square, and the smaller of two large horse barns. The barn had six double open tie stalls and two box stalls. Anyone familiar with barns knows that they can be as cold as a walk-in freezer unless they have a heat source. A crude wing of pole panels was ineffective for getting a cow to the barn. Many times, it required a rope and horse. The only water source was Sage Creek, so we had to chop ice most of the winter. If we had penned cows for calving assistance, we would have to trail them down to the creek to water them.

There was an extensive set of lambing sheds nearby, but they were unusable for cattle. It was a formidable calving challenge to attain much more than survival-of-the-fittest. The following year I convinced Marion and Tom to let me remodel the lambing sheds during the winter to accommodate cows instead of sheep. Being resourceful was a talent I'd inherited from my dad and granddad, who'd lived through the Great Depression. A certain satisfaction comes from using what you have to make something you need.

We used old sheep panels and poles to fashion a usable but crude-looking setup, even managing to design a way to night-calve in the remodeled shed. It was a get-by setup, yet a significant improvement on the old one. We set up a sheepherder wagon next to the shed as shelter and a place to warm up while night calving.

One night while watching heifers, the electric heater quit, so I fired up the Coleman stove to keep the chill off. I was careful to open a window to get rid of the fumes. Later the next day, I started feeling ill, and I was so tired I thought I might be having a relapse of the mononucleosis I'd had in my teen years. After two days, I went to the doctor and discovered I had carbon monoxide poisoning. If I had not been checking heifers frequently, I probably would have cashed it in right there. Almost fifty years later, diesel exhaust smoke will still give me an instant headache.

Our live-calf percentage was much better that year, even better than headquarters. We had quite a bit of sickness in the calves, and I learned a valuable lesson for future use. No matter how exceptional the calving shed and facility may be, it is only as effective as the setup allows for moving the new cow-calf pairs away gradually.

The first few years we were on the Matador, the ranch was breeding the heifers to Jersey bulls for calving ease. The calves were small but not very vigorous. Future attempts to breed some of those Jersey-Hereford cross heifers for cows were unsuccessful. The cattle were very fertile but would not flesh well enough in our environment. The Jersey bulls worked quite well for calving ease at the Texas Matador, and they used Jersey bulls for several years there. The ranch had also briefly experimented with Longhorn and Brahma bulls to increase live-calf percentage on first-calf heifers. Those breeds were not suited to our northern environment and failed to breed up or flesh out.

Until 1974, no fences were breaking up the 80,000 acres of the Sage Creek ranch. With the ranch sitting at a higher elevation, the grazing was more suited to sheep. Sheepherders constantly tending their sheep negated the need for a fence. In addition, as farms and ranches transitioned to motorized transportation and farming equipment, many horses were turned out on the range until World War II, when some were rounded up and shipped to France for meat during the food shortage there. Areas such as Big Basin had been heavily used, especially in recent drought years.

It was pretty devastating when I got my first good look at the entire ranch the following spring. I could stand at one end of Big Basin and count every cow pie as far as I could see to the other side, about four miles away. It was barren.

CHAPTER 5

Range Restoration

"Any healthy system requires periodic harvest and periodic rest." Ray Marxer

The year before I came, Marion had agreed to enter most of Sage Creek into a large-scale *Rest-Rotation Grazing Demonstration Area* designed by August L. "Gus" Hormay for the BLM. It would be one of three major demonstration areas for rest-rotation grazing across the United States and one of the earliest to work with conservation groups in southwest Montana—government and non-government. I felt fortunate to get in on the beginning stages, and we followed the protocol the entire time I was involved with the Matador, almost forty years. Sage Creek was the only ranch out of the original three large-scale demonstration areas that continued the demonstration area long-term.

Gus had a passion for range conservation. While attending college at UC Berkeley, he spent several summers observing grazers, herd movement, and the natural grazing habits of animals across the United States. Gus carefully designed grazing strategies to mesh natural resource use with the natural movements of grazing animals, as opposed to the common trend of directing resource management by attempting to change the natural grazing habits of the animals.

Gus set up exclosures to keep cows out and taller exclosures to keep out all grazers, wild and domestic. He set up a baseline photo-monitoring database, now archived at Montana State University, and conducted occasional tours for observation and teaching.

I learned a great deal from Gus Hormay about range plants and requirements for periodic harvest and periodic rest to maintain health and vigor. Gus designed rest-rotation grazing systems to use a cow's natural grazing habits and movements to achieve that health. At Sage Creek, fences were strategically placed to use the topography and natural barriers to accommodate the movement of a 2,000-plus herd of cows and calves.

The division fences were three-wire suspension fences with posts every sixty to eighty feet, and two-wire stays between posts. Suspension fence had never been tried before in our area and proved to be very effective and wildlife-friendly. Gravity flow pipelines with multiple tanks were installed, making it easier to get cattle to stay in areas that previously had been under-grazed due to distance to water. A side benefit of the new grazing system was improved habitat for wildlife and birds. The newly installed fence and water structures were used to control the time and density of grazing. The one 80,000-acre pasture—about 125 square miles, was split into six pastures. Once we got into the rotation, every year, two of the pastures would be grazed early, two pastures would be grazed after seed-ripe time, and two pastures would receive complete rest from cattle grazing.

Rest-rotation grazing was the primary reason the resident elk herd in Sage Creek almost quadrupled in five years. It grew from fifty-six head in 1974 to four hundred full-time elk by

1979. Winter herds sometimes surpassed 2,500. An interesting observation concerning grazing was that the elk did not go first to the ungrazed, rested pasture with the older, taller stand of grass. Instead, they usually preferred the fresh new growth in the pastures where the cows had grazed early in the season.

Cattle Exclosure monitoring site. Big Basin in Sage Creek. In 1974 this plain was barren due to drought and historically heavy use by sheep and horses. Basin Creek, an excellent Cut-throat Trout fishery is the dark strip in upper third. 2,000 cows rotationally graze the healthy pasture on left. 1998

Most people rely on visual information to comprehend and store helpful knowledge, and I was no different. My past experience with old cowboy ways and ranch traditions—some which were good, and some which needed to change—provided a solid reference point for the new Sage Creek grazing system demonstration area. I had a front-row seat to observe changes in plant and animal health, and the usage that would occur for the next thirty-seven years. I was able to help establish and modify the new rest-rotation grazing strategy that was so successful in restoring and improving this vast prairie called Sage Creek.

We also learned the value of a photographic record of the land and the historical human perspective. My management experience at Sage Creek increased my appreciation and fascination with the ability of this tough country to heal and even flourish. I also began to understand how healthy forage and water are the most critical resources on a ranch. Cattle were designed by the Creator to help manage the grass, the fisheries, encroaching brush, and even fire. As the end product, calves provide economic benefits that allow the ranch business to be economically and environmentally sustainable. Most remarkable is the all-natural conversion of grass into a powerhouse of nutrients available for human consumption.

> *"And I will send grass in thy fields for thy cattle, that thou mayest eat and be full." Deuteronomy 11:15, 1451 BC*

Traditional grazing practices used large pastures and season-long grazing, resulting in some overgrazed areas and some areas left untouched. Both results, overgrazing, and under-grazing, are unhealthy for range plants and have a detrimental effect on the diversity of species and age classes. In the higher-elevation pastures, the presence of larkspur, a purple flower that is common in flower gardens but poisonous to cattle, necessitated that riders continually pushed cows down in early summer. Then in later summer, after the toxic stage had passed, riders would need to keep the cows pushed up higher.

"Riding poison" required riding horseback from the Sage Creek headquarters as far as eighteen miles one-way, every day, plus the miles to move cows, and then the eighteen-mile ride to return home, which added up to many fifty-mile days. I would usually ride one horse out from the barn and lead a second horse. When I got to the cows, I'd switch to the second horse and hobble the first, so he could graze until I was ready to ride home. One horse called Badger, who was inclined to buck when we first got him, got the privilege of carrying me on that big circle without the second horse for eight days straight until he changed his attitude. Work and wet saddle blankets turned him into a pretty good horse and one of the toughest horses I ever rode.

I've got great respect for any horse that can cover four hundred miles, in eight days, in rocky sagebrush country. Badger never got sore. On the contrary, he was in good health and ready to go again. He was the kind that needed regular riding to keep him from getting too fat and sassy.

Even though we'd split the existing 80,000 pasture into six separate pastures, the "smaller" pastures were still just under 14,000 acres each, or about twenty-two square miles in size. Those fenced pastures would provide barriers that kept the cows out of, or in the larkspur areas as needed. Acquiring a stock trailer would reduce the horseback miles, but since we made all cattle movements on horseback, we still covered thousands of miles from the back of a horse.

Eventually, headquarters sent over the 16-foot trailer the ranch had purchased for Sage Creek. We used it to get back and forth to some of the far-flung corners of those 80,000 acres and to get to McKnight Meadows on the other side of the Knox place. We now had a way to haul animals to headquarters or get our horses to the Centennial Valley when helping the crew. The trailer was a gigantic step forward from the two previous decades.

From 1975 to 1980, my Sage Creek crew consisted of myself, my then-wife, and one other young cowboy. Most folks now could not comprehend the miles covered on a large spread by horseback—by every full-time cowboy and cowgirl on the entire ranch. If a big outfit cowboy rode thirty miles per day, one hundred days of the year, he would cover 3,000 miles per year. As of 2022, I still ride a saddle custom-made in 1979 at Three Forks Saddlery. I conservatively estimate that I have ridden that saddle more than 130,000 miles. That is like traveling around the earth's circumference five times on the back of a horse.

Besides the cow herd at Sage Creek, we also ran 400 to 600 yearling heifers or steers at the McKnight Meadows or on top of the mountain in the Ben Holt country. The Ben Holt sits at about 9,000 feet in elevation and is accessed through the neighboring Knox ranch. Driving 400 head of yearlings to the top of the mountain would be the most intense cattle trail each year because it increased about 2,000 feet in elevation in about three miles distance. Before the cattle went up, and while there, we rode pastures to check cattle and gates, and roped anything needing foot rot medicine or other attention. There were always plenty of fences to fix due partly to winter elk usage.

Trailing cattle off the mountain in the early fall was also challenging, especially if there was no snow. In the thirty-seven years I was at the ranch, there were only three years that it did not snow up there by September 12. Snow will usually cause cattle to head for the lower country where forage is easier to find. Otherwise, trailing off a mountain is more challenging than going up since cattle would rather side-hill than go straight down.

Much of the Blacktail side of the mountain had no fence because the steepness of the terrain generally kept the cattle contained. The timbered Jake Canyon made finding and gathering yearlings, the cattle equivalent of teenagers, quite challenging and required several re-rides to locate them all. The main cowboy crew would come to stay with us at Sage Creek for a few days to help wean, and when it was time to wean in the Centennial, I would go to the Staudaher for about a week to help with weaning and shipping there.

Every year we continued to learn more from experience, which improved our efficiency. The more I rode that country, the more I grew to love it, and the stronger my emotional tie to the land became. While many would consider the area desolate, I was fascinated by the somewhat primitive, lost-in-time history and remoteness of Sage Creek. Maybe I looked at it differently because of the many discoveries I would make from the back of a horse.

I loved discovering and exploring old log homesteads that now seem so tiny, almost lost in the sweeping prairies, hills, and mountainous landscapes of the Sage Creek area. Some cabins still had pages of newsprint "wallpaper" clinging to interior logs. I'd often find myself riding horseback over remnants of old stage and freight roads that had once traversed the hills between Salt Lake City, Utah, and the mining camps and early settlements of southwest Montana.

Abandoned sheepherder camps were numerous, and I found and carried home many blackened, shriveled up sheepherder saddles that were decaying in the ruins. "Sheepherder monuments" which were about three-foot-high towers of stacked rocks that sheepherders made, were curious sentinels standing tall on the high, rocky ridges. Poking around in ravine dumps yielded remnants of rusted tin cans, broken enamelware coffee pots and bent plates, pottery shards, and old amber Four Roses bottles.

I especially enjoyed examining flats where I found dozens of Indian tipi rings, usually about ten-to-twelve-foot diameter rings of round stones blending into the natural ground symmetry. Mixed in and around the tipi rings were the accompanying smaller stone circle firepits.

I looked at the area much like the previous inhabitants must have. It had qualities that made me want to be there. Native Americans believed to have been from the Shoshone-Bannock tribe, Sacajawea's people, established large seasonal villages as evidenced by tipi rings still in place after almost one hundred years. The Sage Creek area had diverse plant, bird, and animal species, and abundant trout populations in the small streams.

Numerous game species, including elk, deer, bear, and moose, were likely abundant since they historically traversed the area. The isolated and topographically diverse location would have provided a fair amount of security. For the homesteaders, the place promised opportunity. Out of necessity, homesteaders settled along creeks, or continuous water sources, which explains why most of the private land in the area is on the streams, and the public lands are primarily dry prairie and rocky hills.

These valuable experiences would help me form my beliefs and philosophy about land management. A diversity of plant species and age classes of those species attracts a diversity of animals to the area and also attracts a variety of humans. That philosophy of diversity describes the ranch very well and on a grand scale. Sage Creek is open space at its finest.

The longer I was there, the more the country grew on me despite the rough edges. Or perhaps it was *because* of the rough edges. I was so enamored with Sage Creek and the way my dream was unfolding that I was oblivious to the storm clouds gathering on the horizon.

CHAPTER 6

Upside Down and Inside Out

"Heaviness in the heart of a man maketh it stoop: but a good word maketh it glad." Proverbs 12:25

The year 1980 was both the hardest—and the best year of my life.

We had been at Sage Creek for six years, and even though my then-wife and I didn't exactly share the same values and beliefs, we got along fine—at least I thought we did. Early in the spring of 1980 she went back to California to visit family, and for relief from the long, cold, isolated winter. We were running yearlings at Sage Creek that winter rather than first-calf heifers. The work was much less intense, making it a good time for her to get away.

But weeks turned into months, and she didn't come back except to get her things—and me, if I would go. The last thing I wanted to do was leave Sage Creek and the life I'd always dreamed of and move to California. The unexpected separation hit me like a train, spinning my world upside down as I struggled with the "why" and the "what now?" My life's current three-legged stool—self, family, job—was collapsing. The family-leg had just been kicked out, my self-leg was broken, and the job-leg was pretty shaky. It was the most devastating time of my life as I struggled with the loneliness, the isolation, and personal decisions I was facing.

The single landline phone on the desk in the living room became my lifeline as I spent hours talking to my folks trying to figure out how to fix things. I still had obligations and work that needed done which kept me going during the days. Carl and Shirley Knox, my neighbors ten miles up the road, treated me like one of their own, kindly taking me under their wing and making sure I got some good meals and much-needed company.

As summer got underway, and as work permitted, I rodeoed with Wayne and Olie Else, entering the team-roping event with Wayne. I pushed myself to get out around people, but even among a couple hundred people at the big ropings, I had never felt so alone.

I hated the thought of failure—of divorce—but that was the track we were on. To make matters worse, the company didn't want a single person running Sage Creek. I understood. I'd already seen what could happen when I was helping Gary six years ago. Sage Creek was too remote and isolated and a safety risk for a person alone since there were no communication devices other than the phone in the house. I knew in my gut that moving to California to try and save my marriage wouldn't work. I was pretty sure I'd never be happy again.

I left the television running twenty-four-seven so I would never again have to walk into the deafening silence of that big old empty house. One fuzzy channel came through the rooftop antenna—sometimes two channels. I'd hit bottom. It wasn't the bottom of a bottle, but the bottom of myself. I had nowhere else to turn.

I'd been poking around a bit in an old-fashioned King James Bible Mom had sent me, and it stirred up questions. Jimmy Swaggart had appeared on my TV a few times when I was watching, and I started paying attention. I had serious questions about my spiritual condition. Finally, a different TV evangelist named Lowell Lundstrom, inspired me to take action. I needed to find a preacher I could talk to, so I checked the Dillon yellow pages, and dialed the number of the first church that caught my interest.

An answering machine picked up, so I left my number and a message asking if I could come talk to the pastor. I never received a return call. Next I found a reverend I'd heard of and who was well-liked by several people in town. He answered his phone but refused to see me since I wasn't part of his flock. "You need to talk to your own minister," was his final word. Being brushed off by so-called men of God when I was hurting and looking for spiritual help pretty much convinced me religion was worthless.

So I did something I never thought I would do. I made an appointment with a counselor who had a business in town. She was a very kind lady who listened carefully and asked me some thought-provoking questions. One question probed deep inside. "Ray, you said you've prayed to God. Have you ever asked Him to save you?"

I wasn't sure what she meant. "Well, I guess so...I believe in God, and try to live right," I answered. I'd already told her I was struggling with questions about whether the good things I'd done in life would outweigh the bad things, and if God would allow me into Heaven when I died.

"Believing in God *plus* something you do, is different than completely putting your faith in Jesus Christ and the payment He made for your sins on the Cross," she explained. Then she quoted a Bible verse, James 2:19: "Thou believest that there is one God; thou doest well: the devils also believe, and tremble."

Recognizing my confusion, the counselor quickly quoted some more verses, like Titus 3:5: "Not by works of righteousness which we have done, but according to his mercy he saved us," explaining that there was nothing good I could do to get to Heaven, which is why Christ died in my place. She wrote down one more verse and sent it home with me, saying that I only needed to pray and tell God I knew I was a sinner and unable to save myself, and that I was asking Him to save me the best way I knew how. I appreciated her help in getting me to move my mind to something else, but now I had something else to think about—the free gift of forgiveness and eternal life I only had to accept:

"For by grace are ye saved through faith; and that not of yourselves: it is the gift of God: Not of works, lest any man should boast." Ephesians 2:8-9

On August 18, 1980, I asked Christ to come into my life as my personal Lord and Savior, and guide me through the rest of my life. Without a doubt, that was the single most important, and the absolute best decision I've ever made in my life. From that day on, the joy and productivity in my life took a dramatic upward increase, and I realized, nobody can steal my happiness and joy unless I let them. It was time to move on.

The week leading up to Labor Day is the Beaverhead County Fair, followed by the Labor Day Rodeo—called Montana's Biggest Weekend. Ranch employees were able to take extra time off over Labor Day to go to town and participate in events and festivities before the intense fall work commenced. I always enjoyed helping with the kid's rodeo, the Pro Rodeo, and participating in the wild cow milking event, and different team ropings held in local arenas and at the fair. In 1980 I participated in everything I could. I was hungry for the social camaraderie of the rodeo contestants I'd gotten to know over the past few years, and during the summer when traveling with Wayne and Olie; and for Dillon friends I'd only see on occasion when I got away from Sage Creek.

Life had taken a turn for the better, though I still battled loneliness. I was never a drinker, and disliked bars, but I learned I could still occasionally enjoy the company of friends while sipping a soda, then, since it took over an hour to get home to Sage Creek, I could slip out before things got rowdy.

After Labor Day weekend I had a lot of cattle work that needed done, and no help. Jim and Beth Mooney, some great friends from Dillon, came to Sage Creek and spent a couple weeks with me helping with fall roundup of yearlings and cow-calf pairs for weaning. We got a lot accomplished and had fun doing it.

Jim and Beth were both good hands, and I knew I wouldn't have to worry about them. That didn't mean crazy things couldn't happen though. One day Jim and I were riding a mountainside at Knox's, hunting the last seven or eight head of yearling steers that had evaded our first gather. I was riding a horse of my own, a big Thoroughbred horse called Dave, who was prone to pitch pretty bad if something went wrong. My ribs were wrapped with a big, long ACE bandage. They'd been whacked pretty hard by a scur-horned cow I'd mugged for Wayne Else in the wild cow milking event at Dillon's Labor Day Rodeo, and my ribs were still very tender.

We'd been riding all day and were casually walking our horses down the road coming out of East Creek, following the seven yearlings we'd gathered up. Jim and I were relaxed and visiting as we passed by yet another badger hole in the side of the ditch bank. I guess my horse Dave had gotten bored and was looking for a good excuse to pull the plug, because he saw it too.

Nothing was there, nothing moved, but Dave, that stout sixteen-two hand horse, tall and tough as nails, was sure he saw the boogeyman in that deep, dark hole. He sucked sideways, then made a big powerful jump and went to pitching. Somehow that ACE bandage around my sore ribs got hooked over the whole saddle horn. I couldn't buck off even if I tried. The way he was flopping me around I figured I was about to get beat to death. I managed to keep my stirrups and get him pulled back to a whoa, unhooked my bandage, and we rode off and finished getting the steers through the gate.

I never did get Dave over that counterfeit habit, and neither could any trainers who tried. A year or two later, Dave topped the Bucking Horse Sale in Miles City, and later made the cover of a rodeo magazine in the saddle bronc event. The cowboy was on the ground looking up at Dave's belly. All four of Dave's feet were at least four feet off the ground, and the stirrups were whacking together above the seat of the saddle.

I felt a little better. If he was going to be a bronc, I was just glad he was a good one—for someone else.

The last week Jim and Beth were there, the three of us headed to McKnight Meadows to doctor a few heifers with foot rot. The McKnight is boggy and full of swamp bumps, or hummocks, which required constant vigilance to pick good footing. When doctoring cattle, we'd pick out dry spots where it would be best to rope the animal. We needed the ground to be solid enough that we could get the animal down, then hold them on dry ground.

Jim was riding a bay mare he was planning to sell and wanted to get some more miles and work done on her. She was a pretty nice mare, but she didn't fit what he wanted for a heading horse. Beth and I were both riding young, fairly green horses that were riding well, but didn't have enough experience to handle tough situations. Our plan was that Jim would rope the head on any heifers that needed doctored, and Beth and I would haze it for him out onto dry ground. Then, either Beth or I would rope the heels to get the heifer down and keep the rope tight until she'd gotten her medicine.

We found a big heifer who took off when Jim started building to her, running to the correct position with his rope swinging. Beth and I were on either side, hazing the heifer along to a dry spot. Jim had just caught her around the neck when his horse stumbled, bouncing Jim forward. His mare caught her footing and scrambled back up, but now Jim, who'd bounced really hard, was sitting astride the mare's neck, in front of his saddle horn, with his feet still in the stirrups, the rope slack still in his hand, and his loop still around the neck of the heifer who wasn't waiting around.

Beth and I were riding on either side, glancing at each other silently saying, "this is not going to end well." We needn't have worried. As quick as she got up, that mare stumbled again, and when she popped back up this time, Jim bounced right back into the saddle. He never lost his stirrups or his rope—or the heifer. He dallied his rope, taking a couple of wraps around the saddle horn, one of us roped her heels, and we got the heifer doctored just like trick-riding was part of the routine.

Another great friend who was always cheerful and good about checking in on me occasionally, was Boyd Briggs. Boyd was always up for a good visit. He loved spending time in the Centennial Valley and took me along one day to visit a friend of his in Monida. The abandoned railroad town with one or two human residents occupying some of the decrepit buildings along with the rats and mice and various other creatures, was situated on the Continental Divide near the Montana-Idaho border where we exited I-15 to get to the Centennial Valley from the south side.

The grizzled old fellow we went to visit lived in the upstairs of the Monida Hotel. He did a lot of trapping, and his living and cooking quarters were decorated with numerous beaver castors he'd hung to dry. The ground floor of the hotel housed a small band of sheep. He was happy for a little company, and generously offered to share his stew with us. Fortunately, we were on our way to the valley and had just stopped by to check in on him.

1988 - Range doctoring. Mike Cox with rope, Mitch Stokke, and Bill Whittenberg on calf.

Things were pulling back together in my work life, and since we had a three-day weaning coming up at Sage Creek, I needed to figure out how I was going to keep a hungry crew fed for three days while I was out working. Again, I turned to my family. My mom and my sister, Barb, are both great cooks and were happy to come and bail me out. I deeply appreciated all the help and support I received from friends and family. I had come to realize how much significant personal growth occurs in times of trouble as opposed to the good times.

The roller-coaster year of 1980 was not over yet. I thought I'd experienced the deepest lows, and the highest peaks, and it felt like things were finally leveling out a bit. But once again, I didn't see what was coming at me full steam ahead.

Before weaning started, I went to Dillon to spend some time with another good friend, Cody Brown, at the Club Royal where they usually had live country music on the weekends. I didn't care to drink, but I enjoyed the dancing and socializing. I remember the exact date because it was Mom's birthday—October 18, 1980.

I hated the awkwardness of being single again, but Cody was still single so hanging out with him was easy. We had a small table where we could sit on the fringe and visit while watching people and the dancing. My divorce was over, I was mending, and it just felt good to relax, enjoy myself, and not be worrying anymore.

Across the room I saw Mike Johnson, the local junkman who knew everything that went on in Dillon, hanging around a table with three or four young gals. The band was playing a catchy two-step when one of the ladies, a classy looker, tall and slender with wavy brown hair, got up and headed toward the restrooms in the back. As I watched her, she turned her head toward me—like she was checking to see if I noticed. Oh yeah, I definitely noticed. It wasn't long before she passed by on her way back, and she looked directly at me again. Whoa, what a boost for the wounded esteem! Beyond the smile on my face though, I didn't give it much thought. The farthest thing from my mind was women and relationships.

Pretty quick, Mike came over and pulled a chair over to our table. "Hey Cody…Ray," he nodded in greeting. "You gonna sit there all night?" He jerked his head sideways toward the table with the pretty brown-haired girl. "That young lady on the end is asking who you are, Ray, and if you're single."

My mind immediately forgot that I was not thinking about women as Mike finished delivering his message. "She wants to dance. I told her to just go get you, but she says if you want to dance with her, you'll come ask her." He grinned, pleased with his matchmaker status. I abruptly left him and Cody sitting there. I was on my way.

She smiled when I invited her to dance, and we easily slipped into a smooth, swinging two-step like we'd been dancing together for years. She was easy to talk to and told me she was a civil engineering tech with the Forest Service and had just transferred from Bigfork to Dillon over Labor Day weekend. Her new office friends had invited her to come out with them. She didn't say her name. "How about you?" she asked. "Are you from Dillon?"

I loved how easy she was to dance with and would happily have told her my life story. "My name's Ray Marxer," I began. "I grew up at Ulm, near Great Falls, but I've been here for six years working for the Matador." I wasn't sure what happened, but I immediately sensed a guarded change. "So, what's your name?" I asked, hoping to tilt the conversation back to easy and comfortable.

"Sue."

We two-stepped and twirled around the slick wooden floor in silence. She was smiling, but quiet. The dance was almost over before she gave me my first clue. "I danced with a

Matador cowboy earlier," she told me, nodding toward a cluster of flat hats at the bar. "He was so rude and crude…I walked away and left him standing in the middle of the floor."

Oh boy. I had a pretty good inkling why she'd locked up on me. When Tom Griggs left in 1976, most of his cowboy crew moved on as well. Ron Weekes, who'd taken over as cow foreman, had the big task of rebuilding the cowboy crew. The pool of qualified applicants was never real deep, and at the time, most of the applicants were wanna-be buckaroos who brought a different attitude to the job—and to town.

"I'm really sorry about that," I apologized. "Some of those guys can be pretty rough." I searched quickly for a way to let her know I was different—I really wanted to get to know this girl. "If it helps, I'm not on the cowboy crew. I'm the foreman at the Sage Creek Ranch down by Dell."

That seemed to help a bit because she said, "Well, you don't *look* like a Matador cowboy." Sue continued dancing with me happily and exclusively for the rest of the evening. But she never did tell me her last name, let alone give me a phone number.

CHAPTER 7

New Life—New Job

"...old things are passed away; behold, all things are become new." 2 Corinthians 5:17b

I was not going to give up that easy. A friend of mine knew someone that worked with Sue, and was able to get her number from the third party. I called her and we had a pleasant visit, but when I asked her to go out to dinner with me, she already had plans. I tried again the next week, and again, she already had plans. The third time was a charm.

Sue accepted my invitation for dinner at the Lion's Den. We had a delightful dinner and discussion that evening, that was wonderful for both of us. We closed the place down, just lingering over the cleared table, talking, and enjoying each other's company like long-lost friends. "You know," Sue admitted, "you seemed like such a nice guy until you told me you worked for the Matador." She chuckled wryly, "I heard all about the 'Matador cowboys' from the day I came to Dillon—and none of it was good. Your crude co-worker proved that it wasn't all gossip… like waving a big red flag for me to avoid the whole bunch. I'm glad you persisted."

We quickly discovered that our values and beliefs were alike, we were both from large families and Montana farms—and had both dreamed of being on large ranches. We'd both survived failed marriages, neither of us had children, and best of all, we shared the same Savior. I knew already I'd found my soulmate for life. My long drive home to Sage Creek was short, and filled with wonder, excitement, thanksgiving, and rejoicing.

One long week later, I drove to Dillon to get Sue. We were going to Polaris for a shop-warming dance at the Tash Ranch, and I was excited to introduce her to my friends in the Grasshopper Valley. I debated with myself the entire hour-long drive. Was I going to scare her off if I moved too fast? I was nervous when I came into Sue's trailer to pick her up. I wasn't sure what I was going to do.

Sue was around the corner in the kitchen when I broached the subject. Never one to beat around the bush, it just popped out, "Will you marry me?"

Silence. I stepped around the corner, and she was just standing there with her hands crossed over her heart. Her eyes were wide, and she broke into a big smile. "Yes."

We didn't want people to think we were jumping into anything, so the only person we told was Marion Cross. Marion liked Sue. He was happy for us and relieved that he could alleviate the company's concerns about a single man in the Sage Creek foreman job. I wish I could say that my work never suffered throughout that tough summer, but my heart wasn't in it, and neither was my attention. I've always appreciated my boss, Marion, and the company sticking with me through my struggles. I believe both myself and the company were rewarded for that.

I hired a young kid from Philipsburg who had no experience but was hungry for work. When I taught him to rope the dummy calf head that I'd attached to an old sawhorse, he was a quick study. He helped me feed and knew the routine. I stocked the refrigerator for him, let Marion know I was leaving the young guy on his own, and headed out with Sue on a seven-hundred-mile trip to Canada. It was "meet the family" time.

Sue's parents and four of her younger siblings had moved to Vilna, Alberta, from Bigfork in 1974. It was Christmas, a good time to gift a diamond ring representing infinity. I enjoyed Sue's family, and we had a lot of fun. They had a pretty good inclination of why she'd drug me way up north in the middle of winter. I'd talked to her dad, and in front of the family, on the evening of December 24, 1980, we made our engagement public and official with a diamond ring. On the drive back to Montana, we stopped to see my folks in Ulm so we could celebrate our engagement with them as well.

What a high note to end the roller coaster of 1980! I'd gone from the high of my Sage Creek foreman's job to the depths of an unexpected divorce, which led me to my most incredible high of becoming a child of the King. God knew the whole time what he had in store for me and crowned my year with a wonderful, lovely lady to become my wife.

The way Sue's own story meshed with mine is incredible. She delights in sharing it, and the following are her own words:

"In the summer of 1980, I came clean with God, asking how He could call me His child when I didn't act like it. It felt so good to get a hold of God that once I'd gotten right, I begged God for a husband. 'You must've created a husband for me,' I argued. 'Please show me where he is!' Then I naively gave God a detailed list of everything I wanted in a husband: a tall cowboy, strong, kind, saved, 'who will step on my toes when I need it—but not walk all over me,' etc., ending with a PS: 'And God, I'd really like a husband that loves to dance...unless he likes to dance with other women—in that case, forget it.'"

"Within two months, on Labor Day weekend, He transferred me to Dillon, where I didn't know a soul. A few weeks later, I came face-to-face with my answer to prayer—on a dance floor. Coincidence? Not a chance. The grace of God? Absolutely. We were both young and spiritual babies. But that made God so real to me. Dancing with the husband God gave me, as a very specific answer to prayer, is forever a sweet reminder directly from God of His faithfulness, and how tenderly He cares about every aspect of our lives. God literally answered every single detail on my 'list.' Right down to the optional PS. That still brings tears to my eyes forty-

plus years later and has taught me to pray often, in simple faith, for others."

Sue and I never tire of the joy we find in dancing together. The oneness we share is impossible to describe. Even though it may only be once or twice a year at weddings, special events, or local barn dances, dancing always brings us back to our wedding song, "Could I have this Dance (for the rest of my life)," by Anne Murray.

At this point, Tracy Griggs no longer worked for the Matador, but he and I remained good friends. We team-roped together for a couple of summers in Montana PRCA rodeos. We'd practice in an arena I'd built up at the old sheep pens behind Sage Creek headquarters, where I'd converted the large open holding pen originally used for lambing ewes.

Tracy had met Sue and had seen us together out dancing during our short courtship. He knew me pretty well, and the next time he came to Sage Creek to practice roping with me, he sagely laid out his "Roy Raisin" prognostication:

"Ray Boy," he surmised as we rode back from turning our practice cattle out, "I know what you are going to do. You found the perfect deal in a wife with a good government job that would allow you to rodeo and rope all you want, and you are going to screw it up. You are going to marry her and take her to Sage Creek, and she will quit her job to spend all her time with you."

Tracy had it pegged, for that is precisely what we did. Sue fell in love with Sage Creek for the very same reasons I did. She was ready to be a part of it with me. Tracy and I still managed to hit a few rodeos, and Sue traveled with us that first summer before our family began to grow.

We were married on February 21, 1981, right before calving started. I'd lost my help at Sage Creek, and the feeding still needed to be done. With no assistance offered from headquarters, it looked like we might spend our honeymoon bucking bales and feeding. Again, it was family to the rescue. Two of Sue's brothers, Ron and Fred, who'd come down from Canada for our wedding, offered to stay and feed for a couple of days so we could get away for a quick honeymoon. Ron had to leave, but Fred stayed by himself until we got back, plus a few additional days until we could get him a bus ticket back to Canada. Fred was more than happy to spend a few days learning to rope upstairs in the big barn and go out riding with me. We sent him home with a new cowboy hat.

Sue and I didn't waste any time starting our family. We both wanted children, and Sue wanted to have them before she turned thirty. Clayton showed up on Easter Sunday of April 1982, followed fifteen months later by Kristy in July 1983, and Anna, twenty months after Kristy in March of 1985. Sue liked a doctor eighty miles away in Twin Bridges, and the hospital he used was ninety miles away in Sheridan, Montana.

She never worried about being so far away from the hospital, but I did.

When it was looking like Kristy might show up a couple of weeks early, I hesitated to lead a sizable rest-rotation conservation tour that would take me to the farthest reaches of Sage Creek, but Sue knew how important that was to me and encouraged me to go. Her brother, Dallas, worked for me that summer as did a local kid named Alan Murray. I put both of them to work near the house with instructions about Sue.

Sure enough, just as the tour had stopped for lunch at Teddy Creek, Alan came roaring up in the ranch pickup I'd left behind in case they had to find me. Sue and Dallas, with fifteen-month Clayton, had headed for Sheridan, stopping at my brother Rich's house to drop Clayton off. I made it in plenty of time and ended up having to drive her to Butte over Pipestone Pass in the dark; they were afraid the umbilical cord might be around the baby's neck. Kristy showed up at about midnight, and all was well.

We almost didn't make it with Anna. Fortunately, we were in Dillon when it was time to leave; otherwise, she would have been born at Clark Canyon Dam. For a month, we had three children under the age of three. Sue figured it would be easiest to take care of three babies, get all the nursing, diapers, and teething done in five years, and then have her mobility back. And that's pretty much how it worked out.

Sue rode with me when she could, plus we had two full-time cowboys that she fed three times a day in our home. Every October, during weaning, the entire cowboy crew from headquarters would stay in the old commissary we used for a bunkhouse for three days. Sue would feed an average of twelve men three times a day for three days in our living room while pregnant, nursing, or both. We had to buy all our own groceries, including beef. Sue fed hearty home-cooked meals, including a homemade bread and homemade dessert. Since "board,"—receiving three meals per day—was part of a single cowboy's wages, she wanted them to get their money's worth.

The last year she fed the large crew at Sage Creek for weaning, there was a fire at the Staudaher Cow Camp in the Centennial. The old cookshack burned to the ground, and the cowboy crew no longer had a place to eat. Sue ended up feeding the entire crew for another ten days until we could get an old single-wide trailer into the Staudaher and set up with propane. We used the trailer kitchen as a temporary cookhouse to get through the season at cow camp.

The ranch paid Sue about ten dollars per day per man, and at the end of the year, Sue figured it covered our personal groceries in addition to the crew's. We bought beef by the half. One time, we picked out a big dry ranch cow headed for the auction. We went to the auction and bought her for forty-four cents a pound. We still had to have the cow processed, plus cut and wrapped. She wasn't as good to eat as finished beef, but got us by. The best beef we ever bought came from Ron Johnson's feed yard, where he fed cows fermented by-products from his barley plant. The fat was kind of a yellowish color, but the meat was the tastiest, most tender beef we ever had.

Our closest neighbors were Bill and Bonnie Huntsman and their two children, Mike and Tammi, five miles down the road. Bonnie and Tammi were a tremendous help when we were in a bind for a babysitter. Huntsmans worked hard and long days themselves, plus they spent most of the summer trailing to and living in the Centennial Valley, so we sometimes had to be resourceful in finding ways to tend little ones. Sometimes that involved our personal truck with a topper and bed, and sometimes it involved a pillow over Sue's saddle horn and her right arm for support. I had good help during those years, so generally, if Sue was riding with a little one, it was for the basic need of having the presence of a horse in a gate or bringing along stragglers on the drag. My best ranch horse, steadfast Luke, would be her mount.

Sue's uncle, Doug Wise, gave Clayton a little black pony he named Dandy, when he was two. A year later, when she was ten, Tammi Huntsman had to retire her beloved, gentle "Old Yeller" horse. It broke her heart, but at the same time, Tammi was happy he could be "loaned" to eighteen-month-old Kristy to live out his elderly years. Kristy was eight years old when we had to say goodbye to that faithful Old Yeller horse, who left behind yet another heartbroken little girl.

1985 - Trailing heifers from McKnight Meadows to Sage Creek.
Mitch Stokke on left, Ray Marxer, and Clayton, 3, on Dandy

Sterling Varner, Mr. Koch's right-hand man, who had spent several months on the Beaverhead Ranch when Fred Koch originally purchased it, stopped by our Sage Creek house with Marion on one of his ranch visits and met Sue and baby Clayton. That evening Sue and I had a date to go dancing at the Club Royal. Mr. Varner was there with some others from his group and came to our table to visit. Later, when the waitress came with the sodas we'd ordered, she told us that "that guy over there," indicating Sterling Varner, "opened a thirty-dollar tab for you." Even though we'd probably leave about twenty-five dollars as a tip for the waitress, we appreciated the thoughtful gesture so unique to Mr. Varner.

Sterling and his wife, Paula, were in Dillon when Sue had her first photography show at Western Montana College. Sue was filling in as cook at the Staudaher Cow Camp, and had provided lunch for Sterling's group along, with the cowboy crew, when the photography show was mentioned. Varners made it a point to visit the show, and Paula sent a special thank you note that Sue still has. They were a genuinely gracious, kind, and caring couple.

Larry Angell was a contemporary of mine with the company, having started with the Garvey Ranch in Nevada with his young bride, Susan, in a situation not unlike Sage Creek. Larry eventually became president of Matador Cattle Company, in charge of all the ranches from corporate headquarters in Wichita, Kansas. He would visit regularly, and he also visited our Sage Creek home.

Sue hadn't met Larry yet, and their very first encounter was during weaning when she and baby Clayton were making a quick grocery run to Dell Merc in her little blue Mustang II. Larry was on his way to the pens—the opposite direction—in a small rental car, and the two of them met on the blind corner shortly after Sue had turned onto the main Sage Creek dirt road. They both reacted quickly, missed each other, and went on their ways with elevated heart rates. At lunch, when I introduced them, they laughed and said, "Oh, we already met." Not everyone was as lucky on that curve, and the county eventually fixed it.

In the spring of 1983, the company allowed Sue's twenty-year-old brother Dallas to come to work as a seasonal employee. Through him, we met and hired Mike Cox in 1984, a skinny, totally green young kid with a great deal of untapped talent. There was nothing I enjoyed more than taking a young kid with heart and "want to" and helping them learn. Like Mike, they didn't have any bad habits or claim to know it all. Mike quickly picked up roping skills and developed into a good hand. He was one of the best, most patient cowboys I ever had for working with young horses and working with new broke horses we'd purchased and were integrating into the cavvy.

1989 - Mike Cox on Tommy. Mike McDonald and Kenny Jurenberg wrestling. Jones place. Mike showed Tommy in the Stock Horse show that fall and did very well.

Mitch Stokke from Dillon came to work at Sage Creek while Mike was there. The two of them were good help, whether they were fencing, irrigating, working on waterline, or helping with cow work. They fed off each other's sense of humor and got along great. We enjoyed having them as almost part of the family. They were wonderful with the kids, and made life easier since we never had to worry about the character of the help—who, as always, had open access to our home.

Tragedy struck the following January when we received word that Dallas, who'd gone to work on a seismic crew, had been killed in a seismic accident in Vernal, Utah, at twenty-one years old. Sue was seven months pregnant with Anna. That same year the company announced that they planned to sell the ranches. The Koch family was experiencing family troubles that reverberated through the entire company.

I was restless with the uncertainty of a ranch sale and yet another structural change taking place in Wichita. Sue made a statement to me after she'd spent a few years at Sage Creek, that sometimes working for the ranch reminded her of working for the government because of all the bureaucracy. It was kind of a shocking observation to me since the last thing Koch Industries would want to be compared with, was the government. In the end, she had pegged it pretty well.

I'd reached the peak of my dream of being a foreman on a large ranch. The valuable skills, knowledge, and experience I'd gained while working for the past eleven years made me eager to set new goals and take on more significant challenges in the ranching and the beef industry.

Dallas had reminisced to Sue at the table one day how much he missed going to Sunday school and church as a kid. When he died so young, and unexpectedly, it hit home how important it was to have our own family in a good Sunday school and church. We were praying that God would give us a new job near a good church. We'd had a couple of job offers but did not want to move to Oregon for an excellent job on a well-known progressive ranch—or to a family ranch in Montana that needed what amounted to a hired hand.

I'd brushed off a couple of attempts by the company to get me to move to headquarters and take on the cow boss job since Ron Weekes was leaving. I knew the cow foreman job was the most demanding, most intensive job on the entire ranch. The cow boss was responsible for all the cattle, horses, and range, in addition to cow camps, cooks, crews—and their families. I needed to keep a balance with my own young family and was worried about being consumed with ranch responsibilities. The company finally warned me that if I refused to move into the cow boss job, it would probably be the last time I would have the opportunity to advance my career with the ranch.

Sue wasn't so much opposed to the job as she was to the prospect of giving up the privacy we enjoyed at Sage Creek. She was a bit of a hermit and had no desire to be a part of the hubbub and drama of ranch headquarters. She also had an inner fear that I would no longer need her assistance for cowboy work since I'd now have an entire crew.

As it turned out, God was already way ahead of us. I ended up accepting the cow foreman job, not realizing God's primary purpose for moving us to headquarters was His answer to our prayers for a good, Bible-believing church for our family. The cow foreman's job was the vehicle to provide our living—and for a window of outreach we had no clue was about to open.

> *"The greatest legacy one can pass on to one's children and grandchildren is not money or other material things accumulated in one's life, but rather a legacy of character and faith"* Billy Graham

CHAPTER 8

Cow Foreman—Toughest Job on the Ranch

"Far and away the best prize that life offers is the chance to work hard at work worth doing." Theodore Roosevelt

In December 1985, we finally agreed to become cow foreman on the entire ranch and move to headquarters. It is essential that I said "we" in the former statement. Our goals as a family were the paramount focus of our lives, both presently and in the future. Balance, and the continual striving for balance, would describe our journey into the future. "My" accomplishments would never have occurred without the help and support of Sue and our kids.

Few people have any idea how large the cow boss job on the Matador was. The responsibility included over 10,000 animals—cows, calves, yearlings, bulls, and horses—anywhere from twenty to thirty employees, both full-time and seasonal, a couple of cow camps and the camp cooks, and cowboys' families. The fact that families live on the ranch in company housing, by necessity, is a factor most businesses don't have to contend with and that corporate organizations struggle to understand and appreciate.

In addition to ranch responsibilities, the cow boss was also responsible for taking care of thousands of acres of rangeland, including working with agency people to manage wildlife, fisheries, and hunting season. Over 2,000 miles of fence needed to be maintained, and cattle movements were nomadic, averaging thirty days per pasture depending on the area, season, and water availability. Due to climate and elevation in the arid West, large cattle herds never spend the entire year in one pasture, much like wild bison herds.

When I took over as cow foreman, my family's commitment increased dramatically. In the transition, I was taking on the new duties of cow boss while I was still taking care of Sage Creek. There was a sixty-mile drive between the two ranches. It was a tough few months. A breakout of sickness in the headquarters feedlot didn't help, especially since the cowboy crew was short-handed. We doctored several pens horseback, but we had to run a few entire pens through the chute.

With three toddlers in tow, Sue would keep things going at our Sage Creek home and cook for Mike Cox and Mitch Stokke, our two cowboys. Mike and Mitch were heroes to our kids, and I knew I could count on them to help Sue out if need be. She sorted and packed eleven years of gatherings even though she'd only been on the scene for four of those years. We had all the room we needed at Sage Creek—and then some.

Marion knew she did not want to move to headquarters and made sure that at least we could move into the little house across the creek, which was the most private employee housing at headquarters. He even agreed to replace the living room carpet and build a porch off the kitchen for a freezer, and a place to leave dirty boots, hang hats, and shed outerwear.

Sue was grateful for the concessions and proceeded to haul a few loads every day once the house was ready.

I was responsible for seeing that all cattle work was taken care of and hiring a good portion of a new cowboy crew. Many of the crew that worked for Ron Weekes also left when Ron did, which is usually the case with such changes. Some remained and would become very helpful and essential to the ranch as we advanced. Regardless of who was in charge, I genuinely believe every employee who worked at the ranch made some kind of beneficial contribution.

In addition to Mike and Mitch at Sage Creek, who would help in numerous positions for several years, Dean Deide and Gary Mills stayed on from the former crew and became essential employees. Dean was my right-hand man and helped me accomplish a great deal. Losing Dean is a regret in my career of managing people. I did not appreciate Dean enough or help him enough, but I asked a lot from him. I should have been an enabler to him instead of a boss. Lesson learned, but not always carried out.

Gary and Becky Mills would take my place at Sage Creek and remain there until late 1990. Becky was one of those wives who was good help wherever needed, be it cooking, cleaning, or working the ground crew for branding.

1987 - Dean Deide at the branding pot. Behind him is Dick Slater and Bruce Sandifer. Tom MacKenzie on right.

Susan Marxer photo

In 1986, as the new duties of cow foreman engulfed me, I realized that I had not calved at headquarters or the Blacktail before, but was now responsible for planning and execution of calving. I had to rely on Marion for historical "how to" questions about pasture use and crew placements. We got by all right, but I learned so much about how to proceed in the future. Calving the first-calf heifers—1,200 head—was a nightmare that year as the experimental bulls used on them produced large calves that required a lot of assistance. My crew tried their guts out and got through it. Lesson learned; those employees could not be successful due to someone else's decision the prior year.

The calving of over 4,000 mature cows in the Upper Blacktail went okay. I created a book for the crew to record calving difficulties or tasks they had to perform in each herd and pasture to establish a baseline for the future. We had a crew of six men and a cook living in primitive conditions at the Jake Cow Camp, taking care of the mature cows as had been done for a few decades.

Our records showed after that first year that the herds that got the most attention had the lower live-calf crop percentage at branding. The lower live-calf crop percentage was not a reflection of a great crew of men but a system that was not working. The thousand head of cows in the Wire Field that the crew never rode through, but which only received protein supplement every other day, had the best calf crop. For many years calving had begun between the first and fifteenth of March, so frequent cold temperatures and inclement weather affected the outcomes.

There was a weather bureau station at headquarters that had been there for more than fifty years. I started researching the data collected over the years at that station and found that statistically, there was a fifteen-degree increase in the average mean daily temperature from March 15 to April 1. By changing the calving start date to April 1, we reduced the worst weather-related challenges to calving.

We expected that moving the calving date back fourteen days would push the herd's average birth date fourteen days later. Much to our surprise, the average birth date moved up five days instead. That anomaly was probably because of a longer postpartum period before the bull turn-out date.

The next change was based on the crew's records which showed that we had to milk out several cows at the end of a rope due to a lack of facilities. When the mama got up, she would often be stirred up and run away, leaving the calf, which would go the other way or just lay down. If the cow didn't have good mothering instincts, she would abandon the calf. We did not build facilities. Instead, we tried to eliminate the need to milk out cows by doing a better job of culling those problem cows in the fall. By removing those cows from the herd and introducing a change in our breed makeup, we mostly eliminated the cow-milking chore.

Branding time was very traditional on Montana's Matador and never changed much from the first spring when I learned the system from Cow Boss Tom Griggs. The Flying U brand was used early on, and we later started using the Square & Compass brand on everything, including horses. The Square & Compass was registered by the ranch founders, Poindexter and Orr as the first brand in Montana Territory in 1873.

We worked through one pasture or set of cows at a time, gathering about 300 to 400 cow-calf pairs into a branding trap. The best branding traps were roughly 150 feet square and usually made out of net wire and several stays to keep the wire from bunching. Cattle couldn't crawl through them. The pen size was ideal because it was large enough that we didn't have to sort all the cows off and small enough that the calves couldn't get to running too hard.

We sorted off part of the cows to wait outside the trap. Later in the season, when bulls were in the herd, we would sort off all bulls as well. We usually scheduled brandings three or four days per week, depending on weather and other projects.

1992 Branding - Center back in white hats are Kyle Hardin and Steve Stafford, Middle - standing are Gary and Becky Mills, Front - Sue, Clayton, Anna, Ray Marxer, Tristan Mills and Kristy Marxer

Susan Marxer Photo

The cowboy crew did all branding with horses, ropes, and ground crews. At the earliest brandings, when the calves were small, the ropers necked them, and teams of wrestlers would take them down and hold them. When the calves got bigger, we would heel them and drag them to teams of wrestlers—one on each side of the rope. Learning the correct wrestling technique is of utmost importance and can simplify the job. We would take a break after every hundred calves, and ropers and some of the ground crew would switch jobs.

We cropped the left ear, which was part of our mark, and we used the ear tips to keep a tally of how many calves we branded. We castrated the bulls with a knife, vaccinated them, and in later years, we started ear tagging. Earlier crews had used scoops to dehorn, but we started burning off the horns when I took over.

All the calves were straight horned Hereford bred, so we'd have to dehorn every single one of them. The scoops didn't work very well unless we had a cowboy who had developed solid technique. Otherwise, the scoop either went too deep or not deep enough, and the guys that were good at it wore out doing it. We ended up with a lot of stress on the calf and too much bleeding, so we started doing it differently. I made irons shaped expressly for the job of burning the horn nubs off. First, we'd burn the cap, pop the cap off the horn, then re-burn. The irons were effective, easier to master, and there was much less stress on the calf. After we started crossbreeding with Angus a couple of years later, we had more "muley" calves with no horns, making brandings a little easier and faster.

In the mid-1990s, we started using Nordforks, a metal U-shaped apparatus that is slipped over the back of the calf's neck behind the ears as the roper pulls by. The ground crew member running the fork uses a metal handle for catching and releasing, and a V-shaped rope attaches the headpiece to an innertube that provides stretch and yield between the Nordfork and the heavy stake pounded into the ground. The roper holds the feet tight, and the open-ended headpiece secures the calf safely.

We first tried Nordforks on the bigger calves to save stress on wrestlers and because we started doing a lot more to the calves: more injections, more tags, more electronic tags, implants, and other things depending on their breeding and summer pasture. Nordforks freed up more people, but it was harder on the horses because the horse not only had to drag the calf to the Nordfork, but he also had to turn and face or hold that calf tight until it was done and ready to be turned loose. Nordforks were an excellent way to take care of the calves because they were subdued and working on them was easy. Everybody could do their job quickly, no one was in anybody's way, wrestlers weren't in the way, and the calf was taken care of and turned right back to his mama in about one to two minutes.

Cowboys saved their stoutest horses for the Wire Field calves. Those calves were always the last set we branded, and they came in big and healthy. When the crew finished the day's branding, they'd drive to the Jake Cow Camp, where the cook would have a big meal waiting. The old wooden table and splintery benches outside the small cookhouse would seat the entire crew. Cowboys would wash up in the horse trough or at the pump in front of the Jake cookhouse.

2007 - Sue heeling Wire Field calves on Barney
Photo by Rene Heil

The ranch did everything with their own cowboy crew, with no outside help—unless we were to count family members who pitched in to help with vaccinations, earmarking, tagging, etc.—and cinnamon rolls. The "employees only" requirement was a necessity for safety, but it was one big difference I immediately noticed from the family ranches I was used to, where neighbors helped neighbors. Family brandings were akin to social events, complete with big feasts provided by the family and neighbor ladies after the crew finished the job.

1997 Matador Branding Crew: Todd Sawyer, Al Scafani, Ivan Burch, Dave Mitsunaga, Sandee Nelson, Ryan Kitts, Clayton Marxer, Chris, Ray & Sue Marxer, Vernon Krug - Standing: Rita & Matthew Sawyer, Anna & Kristy Marxer, Robert Sawyer

The Matador had several branding traps, but the main one was the double Tree Culture trap in the Upper Blacktail along the county road. The cow boss and crew counted branded pairs out the gate as they crossed the road into the Upper Bench pasture, which was four or five miles long and was not fenced along the mountainside. There were a couple of little creeks—one on each end.

After the calves had healed up for three or four days, we gathered the herd to the Trail Trap at the southernmost end of the bench to prepare to leave on the trail drive to the Centennial. First, we would hold the herd up in a rodear on the north end where they entered the Trail Trap and work the herd. Cowboys would space themselves around the open, unfenced perimeter of the herd to form the "rodear," keeping the cattle bunched, and allowing the "cut" to pass through as the herd workers brought the dry cows out.

Usually, the cow foreman and manager, Marion Cross, would cut dry cows off the herd. Mostly, we cut out cows that had lost their calf, but we also cut off cows that hadn't calved by branding time. Some of the crew trailed the culled cows to another pasture where they would graze until we'd finished branding the different bunches and had put all the dries together for shipping later.

The rest of the cow-calf pairs stayed overnight in the trail trap, which had a little creek running through for water. We'd get up to the trap about daybreak, open the big gate on the south end and slip in horseback. Beginning quietly at the head of the herd, we'd slowly circle above and below the cows, riding by to get those cows up and alert. We didn't want to move them yet; all we wanted to do was influence them by riding past them to the back end so that the cows would get up, pick up their calves, and mother up—with every mama having her calf at her side. When we got to the back, the whole crew would sit and wait for the cows to find their babies. The bawling would quiet down, and then we'd carefully start working the sides and feathering the pairs out on the trail where they'd line out for miles trailing up Blacktail Road.

June1989, Trailing through Blacktail Canyon on way to Centennial Valley. Herd is strung out beyond bend. Blacktail Creek runs through bottom on right. Pete Slater behind cows.
Susan Marxer photo

The cows usually had figured out that when we started riding back by them, it was time to find their baby and get mothered up because it was time to hit the trail. We'd move along with them past the old Chaffin place and the buffalo jump, trail them through the fenced Anderson trail lane, and past the Landon forty. The cows knew they were heading for summer pasture in the Centennial Valley, thirty-some miles away. Once we got that bunch out of the way and on the trail, we returned and gathered the next set of pairs to the Tree Culture or whatever pasture we would be branding next. The following day we'd brand another bunch, then head up the Blacktail to trail the pairs already on the road.

It was a repetitive process, so every few days, we were branding another big bunch of cattle, or if we were trying to get it done fast, we might go to another area farther down by the Landon and brand a bunch in a trap there. We'd turn that bunch on to the Middle Bench to heal up. We had something going all the time, working 5,000 cows (more in later years) in bunches of 300 to 600 toward their summer country. Some cows would be trailed over the hills into Sage Creek through Red Canyon and then into Teddy Creek. The rest summered in various pastures in the Centennial Valley and surrounding mountain country.

Usually, about six crew members and a cook lived in the Jake Cow Camp until most of the different bunches had been trailed up the Blacktail to the various summer pastures. Around the Fourth of July, the single members of the cowboy crew and the Jake cook moved up to the Staudaher Cow Camp, where they would live until early winter. Once we'd finished weaning and had trailed the cows and horses back to their winter pastures in the Lower Blacktail, the crew moved back to headquarters, and the cycle started over.

Some of the married cowboy crew might be requested to remain at headquarters for part of the summer to help hay. Often, cowboys were not interested in riding tractors and were likely to seek employment elsewhere rather than live at headquarters and farm. The living accommodations in the valley were sparse in quantity and quality, many with no modern conveniences like power or running water.

The cow work was probably pretty typical for big sprawling western ranches. The cows were unlikely to see a corral more than twice a year. Once was during fall weaning when we gathered them to the Sage Creek corrals or the Staudaher pens, where we would wean the calves off the cows and ship the calves in trucks, then turn the cows back out to graze until winter took hold.

The other time cows saw the corral was later in the fall in November or early December when we trailed them home to the Jake on the Blacktail side or to the Sage Creek headquarters for pregnancy testing and to give them their vaccinations. We peeled off any open cows (not bred) and ran them in a separate pasture so we could take them to the auction. We turned the bred cows back out to graze on pastures we'd reserved through the summer. We did all our own preg-testing.

In 1975 Tom Griggs, Ron Weekes, and their families lived at the 7L Ranch on the south side of the valley in the old guest ranch buildings. It was too far from the Staudaher and most of the work, and the flies and mosquitoes were horrible. Even though they had electricity and running water at the 7L, they moved back to the Staudaher Cow Camp the following year.

The Matador was usually "buyer of choice" when neighboring ranchers retired, and the ranch eventually added about 100,000 acres to the original 240,000 acres—including the attached BLM and state lands. Following the purchase of several adjoining ranches, the housing would change 180 degrees. Now, there is more housing than crew. One thing that has not changed is that from December to mid-May the Centennial Valley is a great place to be from due to the severe winters.

November weaning at the Staudaher pens in Montana's Centennial Valley. Shane Lord, Kyle Hardin, Ray Marxer, Sandee Nelson.

CHAPTER 9

Cow Camp Summers

"Don't judge each day by the harvest you reap but by the seeds that you plant." Robert Louis Stevenson

The old cookshack at the Staudaher Cow Camp had burned to the ground the fall of 1985 when the crew was weaning at Sage Creek. Marion and I, with George Munski, had finalized plans for a new combination cookhouse-bunkhouse. The building was an eighty-foot by thirty-foot rectangle with a wall dividing the two units. We decided the most durable and easiest-to-maintain structure would be a metal building on a concrete base that would serve as the floor.

The large room on the north end would hold bunks and sturdy hangers for the crew and have the first indoor bathroom and showers in the Staudaher's history. The south end had a bedroom and bath for the cook, and a full kitchen and dining area took up the entire center section, sharing the wall with the bunkhouse. Interior walls were Sheetrocked, taped, and painted. We installed propane heaters and lights, a propane stove and oven, and two propane refrigerators—one of which would serve as a freezer. George Munski did a lot of the work and planning on the new building. He was probably the handiest jack-of-all-trades that ever worked on the ranch.

George worked for the company for thirty years before retiring.

We also had a well drilled and installed a pull-start gas pump to pump the water to a 1,000-gallon water tank on the high hill behind the cookhouse. The water gravity flowed to the indoor plumbing of the cookhouse-bunkhouse and to the two single-wide trailers we set up that same spring. The cook would usually fire up the pump after the crew left and fill the tank on the hill daily. Sometimes we would fill it a second time if needed.

The trailer they'd brought up the previous fall for a temporary cookhouse was moved and set up for family use. We installed propane lights, heaters, and a propane stove and refrigerator. I got permission from Marion to take the oldest of the trailers from headquarters to the Staudaher for my own family. That trailer was installed across the camp from the first one and converted to propane. In addition, there were two or three camp trailers for couples, with enough room for one or two small children.

The cowboy crew had progressed to having two crew-cab pickups and two twenty-foot horse trailers, which meant the crew still mostly went everywhere together. Families who didn't want to wait on the cow boss or an employee for a ride when they needed groceries, had laundry to do, kids in school, etc. had to drive their personal vehicles and provide their own gas. Some families found it easier to stay at headquarters ten miles from town. It wasn't easy at times, yet it was enjoyable for most. Not everyone appreciated the rules of conduct I established, such as no alcohol in camp. Still, rules were necessary to prevent problems for a group of good folks working together all day, then living in the same space the rest of the time.

Cowboys had saddled horses by lantern light since the original construction of the barn around 1910. The aged, dry wood combined with years of accumulated floor duff and dust was a significant fire hazard when combined with lanterns. I ran a simple electrical wire around the work area so we could install electric lights run by a small portable gas generator. A cubby box that I built on the front of the barn protected the generator and made it easy to access. Sometimes Sue and the kids would pull their little red wagon to the barn and haul the generator up to the trailer house so she could vacuum the dingy, old, sculptured carpet.

Dean Deide, Dick Slater, Don Reese, Clyde Kenyon, Craig Filmore, and Rod Richardson were the cowboys who moved to the Staudaher that first summer of 1986. One or two left, but the rest all pitched in and helped fix the barn stalls and add hangers and saddle racks. They also patched the worst holes in the steep barn roof and helped build an arena beside the barn corrals, complete with a rusty old, modified chute. Sue and I leased a few roping steers so anyone who wanted to work on their roping, their horses, or have a little fun would have a good way to fill the evenings in camp.

Kristy's Old Yeller horse, Sue's colt Koko, and Happy Appy the wrangle horse roamed around camp as our lawnmowers and spent a lot of time at our trailer, where the kids spoiled them with handfuls of dog food. One night we heard a loud crashing sound in our crude narrow entry—where we stored the dog food. Old Yeller had come right up to the door to help himself. After that, we had to keep the dog food in our horse trailer beside the house.

When we could take the kids to work with us, we did. Other families with suitable kids' horses and supervision were welcome to do likewise. I often took four-year-old Clayton horseback with me, where he would ride for entire days with the crew and me. He loved cowboy work, even coming into our room at 2:30 a.m. one morning, fully dressed, wondering when we were leaving. When Sue was helping, Kristy and Anna would ride too. Kristy rode Old Yeller, and Anna rode Alpo—the original horse Tom Griggs cut out for me and who was now retired.

1987 - Ray Marxer moving small bunch of cows in Centennial with 2-yr old Anna, 3-yr old Kristy, and 5-yr old Clayton--and Sue.

The kids learned to entertain themselves by playing around camp in the little creek or on the hill overlooking cowcamp, but mainly in the big old historic horse barn. They loved "training" the old wrangle horse Happy Appy—bareback with a halter—by trying to get him to jump the little creek in front of the barn. Happy would always take an extra little hop or two, and the kids thought they were pretty punchy riding a "bucking horse" bareback.

As a family living in a remote location, we made our own entertainment with once-a-week storytelling time. Each of the kids would take their turn at making up a story and sharing it with the rest of us. Sue and I told stories as well. They all were good at using their imagination and sharing their stories. Anna was exceptional with her fantastic imagination. She loved telling stories and would gladly take another's place when they had a brain cramp. Purple Sharkey, her imaginary friend, always came up with something.

Other times, the whole crew and families were invited to Lakeview to swim. The Rush family had the Lakeview Guest Ranch, and they would grill burgers and host us for an afternoon of swimming in the indoor Alpine Pool.

Sue and I and the kids took a pack-trip up behind camp in Peterson Basin one year. While setting up camp, I heard Clayton yelling, "Dad! Dad! There goes Jake!" I looked up just in time to see our packhorse, Jake, disappearing into the trees and heading back to the barn.

I jumped on my horse bareback, thinking I'd catch right up to Jake. But Jake led me on a merry chase clear back to the gate we'd fortunately, closed behind us. We got caught in a storm coming out the next day and looked like a bunch of drowned rats when we got to the barn. That trip is still one of the favorite memories of our cow camp years.

We were all so fortunate to live in the Centennial Valley. Even though we were only about forty-five miles from West Yellowstone, we seldom went to the park. We could see all the same animals almost daily, except for bison and humans, and never leave the ranch. I am confident our family would all say those five summers at the Staudaher Cow Camp were some of the best in our lives.

We had good crews through those years, but still experienced the typically high turnover rate that would plague us until we changed the structure of how we managed the ranch in the early 1990s. Bruce Sandifer and Dick Slater would spend a lot of time with me while they were there. Many of our experiences included four-year-old Clayton horseback with us.

Another solid old-time cowboy was Jim Bishop, a hand that was slightly older than I, that would work for us for several years. He was so good to our kids and took Clayton on horseback a great deal—especially working the sides of a trail herd. While Jim would humbly argue he was not the greatest cowboy, he was, in fact, a very good one. Jim was like a methodical machine when heeling calves in a branding trap. He knew how to string out a herd, and he had a unique ability to start a large herd of pairs on a trail and have them mothered up better than anyone I have seen.

I would estimate that Clayton probably had ridden 5,000 miles by the time he was nine or ten. It was not unusual for him to be following 500 or 600 heifers by himself down the South Valley Road while the crew and I were gathering more from the mountains to put with them. Our good neighbor, Ed Wolfe, frequently visited with him and offered him sandwiches and sodas while he trailed those heifers alone.

Clayton, 7, and Alpo finished trailing and are waiting at the Jake gate for a ride home

Sue and the kids usually followed the herd's drag up the Blacktail while the crew and I were miles ahead with the main herd. The drag would be cattle that had run back during the night to where they'd started or some that never moved on with the herd. Usually, they would pick up a mis-mothered bunch of cows, several calves, and up to six bulls which sometimes, cowboys would purposely drop because bulls were a pain to trail. Before the summer was over, Sue and the kids would have ridden the entire length of that sixty-mile trail behind cows. Sue sewed denim jean-leg lunch bags to carry over their saddle horns.

Kristy and Anna rode many miles as three and four-year-old tots and many more through their high school years. Most miles were behind cows, but they also rode with us to look for cattle on the Continental Divide Trail along the Montana-Idaho border, in the Blacktail, and the US Forest allotment at Long Creek.

The kids grew up in the most fantastic natural science lab—God's creation. They'd collect white terrestrial snail shells they discovered on top of the hill behind camp, convinced they were left there from Noah's flood. They watched calves being born in the pasture and observed cowboys assisting with a difficult birth. They bottle-fed bum calves and bum lambs that trucker Roger Cleverly would give to them. They cleaned pens with a pitchfork when they were old enough to spend regular time at the calving sheds, and once we moved to headquarters, it was their job to clean the horse barn.

We were fortunate to provide our three kids with a unique upbringing uncommon in the latter part of the twentieth century. They were exposed to many different people and developed the ability to see people for their character and work ethic. At a young age, all three kids could tell a bullshitter from someone genuine.

"In all labour there is profit: but the talk of the lips tendeth only to penury." Proverbs 14:23

Another quality gained from their upbringing was approaching their future jobs with an owner's attitude and happily doing whatever was needed to accomplish the job. Like the cowboys, our favorite season on the ranch was branding and involved our entire family from the time the kids were preschoolers. We branded over 6,000 calves every year from early May until the 4th of July for twenty-five years.

Occasionally we'd reward the kids with a ten-dollar bill for the Labor Day carnival, but mostly the reward was in the work. It was our life, and we enjoyed what we did.

When the branding finished about the 4th of July, the crew was tired and deserved some time off. There were still pairs needing to be trailed thirty-five miles to the Centennial Valley. It had become a family tradition for Sue and I and the kids to trail some of the last herds while giving the crew four or five days off over the holiday weekend. When they returned, they would immediately move to the Staudaher Cow Camp.

Sometimes we'd trail all day and then tent camp behind the cows at night. A side benefit of camping behind the cows was that we didn't have much trouble with stragglers going back during the night. After breakfast, the kids and I would start the herd back up the road while Sue packed up camp. She'd bring the truck and trailer to catch up and help get the cows to their destination.

July 4, 1992, Clayton on Frank, Kristy on JD, and Anna on Peter Paint. Trailing herd to Centennial Valley summer pasture with parents, Ray and Sue Marxer ©*Susan Marxer*

Blacktail Canyon was a favorite for the kids. A cave near the top provided fuel for wild imaginations, and they explored that whole stretch of cliffs and trails while I fished Blacktail Creek. Sue hiked down into Anderson's meadow to poke around in the remains of a log homestead cabin and discovered a big patch of wild strawberries ripe and ready to pick. She filled her hat with enough for breakfast, then we stopped on our way home so the kids could help pick more to take home for wild strawberry muffins. Another typical day off in the valley for our family was looking for arrowheads and buffalo skulls.

In 1986 or 1987, the ranch purchased the ranch of Bill and Eileen Jones on the south side of the valley. Bill and Eileen continued to live there for a couple of years, but the single-wide trailer and cabin were available for a summer residence for a couple of our crew. One of the most significant benefits was that we now had a phone and an electric freezer only twelve miles away.

Eventually, Bill and Eileen Jones retired, and some of the crew would take responsibility for the south valley area and live in the main house. Some of the families who lived there were Todd and Rita Sawyer with their two young boys, Robert and Matthew, Rod and Audrey Wipperling, Justin and Andrea Davis, and Dave and Jessica Nick. Sandee Nelson, a long-term cowgirl on the crew, lived in the trailer for several years. Jim Bishop lived in the cabin for a while, and we also kept an electric chest-type freezer there for cookhouse beef, and for ranch families to use.

The cow boss years would be significant in our family's journey. While it was the formative years for our three young children, it was also a time of spiritual growth. God was so good to us in placing us in a church that preached and taught the pure Word of God. The Bible came alive for both Sue and me. I am so thankful for her commitment to visiting different churches to find the right one while I was in the middle of calving season.

Once exposed to some of the truths of the Bible, we were hungry to learn more. It had been years since I had regularly attended church and rarely as an adult. After attending our chosen church a few times, I was asked by a friend what I thought about it. I still remember my genuine but ignorant response. "I like the preaching, but the singing is a little too gung ho." I'd never heard singing like that in church, but it didn't take long before I developed a love for the old-time Gospel hymns and the enthusiastic spirit of our congregational singing. We soon developed a habit of being in church whenever the doors were open.

During the summer months of July through November, when living in the remote Staudaher Cow Camp in the Centennial Valley, we had no power, no phone, no mail, and no distractions. The propane lights heated up the trailer so bad in the summer that we went to bed with the sun—what a perfect situation for spending quality time with our three little kids. We read them a lot of books, including the Bible. I read the King James Bible from cover to cover for the first of numerous times in the years to follow. I listened to an entire box of cassette tapes of old-time preachers. Oliver B. Green and B.R. Lakin were my favorites.

Regular church attendance was out of the question living in the hills fifty miles from town. Our pastor recognized the challenge and helped us turn an obstacle into an opportunity. Every Friday in mid-afternoon, he came to the Staudaher and held a Bible study for anyone who wanted to attend. Some were Christians, and some not, but most came. They got to hear the Word of God, and some received Christ as their personal Savior, and some did not. Nothing thrills our souls like running into some of those cowboys many years later and hearing their profession of faith that started with seeds planted at those cow camp Bible studies.

Marion's wife, Jackie Cross, ran the commissary for the ranch and filled grocery lists for the cooks. Jackie timed her deliveries each week so she could be at the Bible studies. She had received Christ following a Bible study at our home and would be a faithful follower until her passing many years later.

While a cow camp ministry was unique for us, it was also memorable for many great preachers and evangelists who would accompany our preacher. I remember Ken Stertz, a Bible runner to eastern Europe; Ron Sykes, an ex-biker who became a missionary in South Africa; Sam Gipp, an evangelist; Tom Williams, an evangelist who loved roping calves, and others. We heard preaching most of us would not have heard otherwise, and the cow camp experience was a blessing to the preachers. I am humbled every time I think about the relationships God has allowed me to have with numerous great preachers and their families. Those Bible studies and spiritual influences have had a profound and lasting impact on our entire family.

When it was time to begin schooling, we decided to enroll the kids in a private school using ACE, Accelerated Christian Education curriculum, which could be accredited. Clayton and Kristy started kindergarten the same year, so Sue moved to headquarters after Labor Day weekend that year and the following year when Anna began. After that, Sue was able to take their lessons to the Staudaher Cow Camp and homeschool until we moved back to headquarters in mid-November.

We began homeschooling using the same curriculum in the middle grades, except we switched to Saxon math. Every year the kids took achievement tests to ensure we were on track. We homeschooled through high school, and Clayton and Kristy took college ACTs, with Clayton scoring in the top 20 percent and Kristy in the top 5 percent of college-bound seniors in the nation. Anna didn't go to college but was selected as the public relations officer on the Montana 4-H Ambassador Team and spent a year traveling from coast to coast, speaking and participating in leadership and events. The summer Clayton graduated, he had the opportunity to travel across the country with our Horse Clinician, Mike Bridges, and his youngest son Justin, learning and assisting with clinics. Kristy who was planning on going into Dental Hygiene, spent the summer working with our family dentist who also happened to be a great neighbor.

Robert Sawyer is another terrific ranch-raised kid. When I hired his dad Todd for the Matador cowboy crew, Robert was only a baby. After the ranch purchased the Jones place, Todd's family lived in the old brown house in the Centennial during the summers. When the ranch bought the old Chaffin place in the early 2000s, Todd, his wife Rita, and their two boys, Robert and Matthew, lived there year-round—Todd and Rita still do. The Chaffin place, which we called the Price Creek Division, was twenty miles up the Blacktail from headquarters, and out of range for electricity and phone.

2009 - Matador's Price Creek division, the old Selway place. Ray Marxer, Jake Trindle, Todd Sawyer, Jake Butte, Sandee Nelson.
Susan Marxer photo

The huge three-story salt-box style house was originally built as a bunkhouse and cookhouse for the historic Selway ranch. We found an old newspaper clipping detailing the house-warming for what was considered state-of-the-art housing early in the 20th-century. About 300 folks from the Dillon area traveled the thirty miles up that rough dirt road horseback, and in wagons and carriages to attend the three-day celebration. The largest barn was three stories high and possessed an elevator. There were giant grain bins on the third floor. A sheep ramp went from the ground floor to the second floor which was set up for lambing ewes. The house was set up with banks of batteries that provided adequate electricity, and we got Todd and Rita a satellite phone which was pretty hit and miss, but was better than nothing.

Rita would drive their two boys to the school bus and pick them up every day at Brown's Lane—two twenty-mile round trips each day. In between school-bus runs, Rita still managed to work the brandings and irrigate the Upper Blacktail meadows, improve the landscaping, rebuild the kitchen cupboards, and hitch beautiful horse-hair pieces. Robert went to school back east and was accepted into an Officer's Training Program. He became a fighter pilot, landing and taking off from aircraft carriers at sea.

2005 - ranch kids done branding. Matthew Sawyer, Anna Marxer, Robert Sawyer, Kristy Marxer

Robert is still that humble, good-natured kid who grew up in the freedom of a remote ranch. He and his younger brother Matthew, who prefers hunting and fishing, remind me of a modern-day Tom Sawyer and Huck Finn.

CHAPTER 10

Computers, Spreadsheets, and Crossbreeding

"Vision animates, inspires, transforms purpose into action."
Warren Bennis

Beginning in 1986, the company committed to the foremen and managers of the ranches to help us learn new things that could benefit us, the ranch, and the company. We were exposed to personal computers for the first time and how to use them; some never did. I still recall the paper handout they gave us as we sat before a computer for the first time at a training in Wichita headquarters. The first page had a picture of a PC and, in big, bold letters clearly stated, "THIS IS A COMPUTER." They certainly knew where to start.

Shortly after our PC introduction, the company exposed us "cowboys" to some very new ways of approaching business. Our first lesson was William Edwards Deming's approach, often referred to as "Continuous Improvement," which involved, among other things, statistical process control (SPC). We learned how to use computers and charts to visualize and analyze our business. Pareto charts would help us visualize how our actual costs compared and where we had the most opportunity for improvement.

We soon learned that non-grazeable feed cost (substitute feed or supplemental feed) was the most considerable cost by a significant margin. Labor was a distant second, and depreciation was the third greatest cost. The rest of the costs were much smaller but still significant. These highest costs are still the same in the livestock ranching business, regardless of operation size or regional location.

We did a very detailed analysis of our operation enterprises, which provided the facts needed for us to improve. Individual operations in our company structure in 1988 had not changed much from 1974 except that we had sold the Roberts' "Bug" Ranch in Wyoming, and the lease was let go for the Garvey Ranch in Nevada. The remaining three ranches continued operation until 2021—the Matador Ranch in Matador, Texas, Spring Creek Ranch in Eureka, Kansas, and Beaverhead Ranch in southwest Montana.

We ran a straight Hereford cow herd. From personal experience on a much smaller scale, I was confident we could improve our beef product with crossbreeding. Marion and others were quite reluctant to introduce a breeding cross since they knew how important it was to have a cow herd that fit our environment. A cow herd that could graze most of the year—if not all year—and still perform was critical. For many years, the herd that had successfully done that for the Beaverhead was straight Hereford breeding. Our ranch cows averaged about 1,080 pounds, depending on the forage year.

Most human beings are good at seeing how we think things could be better but poor at accomplishing it. My desire, and later, my talent, was to figure out how to make something work for us that was a perceived obstacle. Marion's concern was that if we crossbred some of the herd, we would soon run out of Hereford cows that fit our environment. Challenge accepted.

> *"If you are working on something exciting that you really care about, you don't have to be pushed. The vision pulls you."* Steve Jobs

With my minimal exposure to computers and spreadsheets, I launched off into my career's first significant learning project. Using the multi-plan spreadsheet that was available at the time, I decided to create a spreadsheet that would include every aspect of the cow herd:

- inventory by age
- cows
- bulls
- pregnancy percentage
- live-calf crop percentage
- cull rates
- calves produced
- heifers retained for the cow herd

Then using the rates and percentages that Marion had recorded over time, I built a massive spreadsheet that was approximately eighteen inches wide by twenty inches long when printed on one sheet of paper. That was our baseline. The power of computers quickly became evident to me when I learned how to copy that info into another spreadsheet with formulas that transferred the first year's results into the second year, then into the third year, and so on until I had extrapolated ten years out.

The calculations from those results allowed me to demonstrate to Marion that by breeding 2,400 cows per year to Angus bulls, we would still have a certain amount of Hereford cows left as our base. That spreadsheet, and conversations with Larry Angell and Bob Kilmer, convinced Marion to try crossbreeding. Trying those Angus bulls crossed with the first set of our Hereford cows worked even better than expected. Crossbreeding improved reproductive performance, and we were able to reduce cull rates substantially due to fewer eye and udder problems. Lack of skin pigment in the Hereford's white face and udder often resulted in sun damage, which was worsened by reflection off the snow in the winter.

1986 - Marion Cross, Larry Angell, Bob Kilmer. Sign on left has wagon-rod sheep brand, and square and compass cow and horse brand. Photo on wall shows Marion preparing Alpo for young Chase Koch to ride during a visit to Montana's Matador Cattle Company. *Susan Marxer Photo*

With the experience of building that spreadsheet, I realized how similar use of the computer could help us. We could experiment with "what if" questions and see the outcomes, which would be monumental for me in decision-making for the rest of my life.

I thought about the next "what if" question. "How would these crossbreeding changes affect our product and profitability?" So, it was elementary for me to add some columns and lines to the spreadsheets that would include pounds of weaned weight per calf produced. I sought advice from neighbors and other producers who had experience with Hereford-Angus cattle to establish reasonable expectations. I went even more conservative and estimated that the F1 Black Baldy 205-day calf would be twenty-two pounds heavier than their Hereford contemporaries with the same inputs.

When I saw the increase in pounds alone over ten years, I was shocked at what we had left on the table for so many years. After implementation, the actual difference in weight was thirty-four pounds. The big shocker that allowed us to make a breakthrough change was when we got the resulting Black Baldy heifer into the cow herd. Her calves weighed seventy pounds more than the Hereford heifers bred to the same bull.

One bit of caution when crossbreeding—while heterosis is a great way to improve, it may also increase cow size, possibly detrimental to operational cost and performance. Having a cow herd that fits the environment is still one of the most critical drivers of profit in the beef industry. It will remain that way over time due to the unique ability of a cow to convert what grows on the earth's non-tillable land into something that is life-sustaining for humans. (An entire chapter, *Essential Partners—Our Cattle,* later in this book, is devoted to our breeding program, and maintaining optimal cow size in our herd.)

That spreadsheet project for crossbreeding projections established my analytical and reasonable approach to future changes and opportunities we would consider. I consider that project a "master's thesis," a template for significant improvement in ranch management. I continue to use variations of my original analytical spreadsheet in our consulting services.

"Ideas are funny little things. They won't work unless you do." I don't know who said that, but it's true. Anyone can come up with ideas, then expect someone else to put them into practice. Don't just come up with ideas. Flesh them out and show expected outcomes using best-known facts and figures. The old cliché, "In God we trust, everyone else bring data," works.

The last few years of the 1980s brought about numerous changes—the increase in the size of Koch Industries and an increase in government regulation influenced many of those changes. A new law was the I-9 proof of citizenship form, which all employees had to provide to prove they could legally work. It served a purpose but was a duplication of information for veterans and long-term employees. Imagine my frustration and embarrassment when I had to go to Roy Drinnen, our ninety-year-old fencer, and ask him to fill out proof-of-citizenship papers. Roy was a World War II veteran who had served his country on the beach of Normandy on D-Day, and now I was forced to ask him to come up with proof of his citizenship. This subject still fires me up! We have to prove citizenship to work in this country but not to vote legally. Wrong. Wrong. Wrong.

Roy Drinnen herded sheep for Jack Thomas, our neighbor in the Centennial, for twenty-eight years before he came to work for the Matador at the age of seventy-five. Roy tended 2,000 miles of fence for us, walking every mile and shaking every post. He refused the offer of a four-wheeler or a helper. We had some federal-lease pastures that were off-limits to vehicles. Roy would walk in, sometimes ten miles carrying his tools and packing posts over his shoulder.

Roy Drinnen, Matador Fencer

He drove his own Dodge pickup and pulled his little camp-trailer, beginning when the frost went out of the ground in the spring until late fall when it froze up for the winter. A motel room in Dillon or his own winterized camper would be his home for the winter. The following spring, he'd begin in the Lower Blacktail as soon as the frost went out of the ground and work his way upcountry, trying to stay ahead of the cows we'd be trailing.

I can't say that we never worried about Roy just a little, and there were probably some who thought we were taking advantage of an old man. Roy didn't think of himself as old, and he loved the opportunity to build the stoutest fence in the country. I made sure to look him up and spend time with him regularly. I was the only one he would allow to sharpen his digging bar, so I saw him often. He was fiercely independent, and enjoyed his job and being out in the hills with Nellie, his faithful dog. I always knew there was a chance I might find him deceased, which I dreaded. But I also knew he was happiest making his own decisions, even if that meant passing peacefully in his element. Thankfully that never happened, but Nellie died, which was hard on Roy. At age ninety-one, Roy could no longer see good enough to renew his driver's license. He knew it was time to retire, and he moved into assisted living in Dillon.

An important position for the ranch during these years was the camp cook. We had a full-time cook at the headquarters' cookhouse, which had all the modern conveniences and a decent living quarters upstairs. The farm crew ate there, and full-time single cowboys ate there during the winter. The camp cook needed to be hired every spring since the job was seasonal—mid-March through July fourth, at Jake Cow Camp, and about July 5 through mid-November at the Staudaher Cow Camp.

A large part of compensation for single employees was "board"—having their meals provided. If the cook was good, their wage was fair, but if the cook was poor, so was their compensation and attitude. Cooking for crews with the lack of modern conveniences was difficult at best and resulted in a frequent turnover. We went through five cooks the first year I was cow boss, including Sue since she filled in between cooks.

Later, we found an incredible cook who, at one time, had owned a five-star deli near Seattle. He was a gentleman by the name of Tom Serginson. A great guy who was easy to get along with and performed magic in that tiny homestead cookhouse with the water pump out front. Tom could provide a great meal regardless of crew size. He made treats for branding and turned out fancy cream puffs in that primitive kitchen. Often the Jake cook would be done when it was time to move to the Staudaher, and we'd have to find a new cook. Tom moved with us and cooked at both camps. He was kind to the cow camp kids, and he let Sue put her meals in the cookhouse oven with his to avoid heating our trailer worse than it already was. Sadly, Tom struggled with alcohol. After a couple of mostly dry years, he moved on and was sorely missed—and never replaced. We were fortunate to have had Tom.

Matador's Jake Cookhouse
Best cook ever, Tom Serginson w/
Susie, and great hand, Jim Bishop
©1988 Susan Marxer

Cook stories could fill a book by themselves, but not this one. When we restructured the ranch management in the early 1990s, I closed the cookhouses and we no longer provided employee meals as part of employee compensation. We added cooking facilities in the Sage Creek bunkhouse and in the headquarters bunkhouse so single employees could receive full pay and manage their own meals. Thus ended a long-standing tradition and a constant breeding ground for discontent on the ranch. The reliance on large crews of primarily single hands had transitioned to smaller crews of mostly family men.

The late 1980s and the 1990s would see the most dramatic changes on the Beaverhead Ranch during my career. Much of the change was due to the remarkable increase in size and scope of business for Koch Industries, Inc. When I started with the company in 1974, I believe there were around 1,300 total employees in Koch. That number would expand to over 60,000 employees by the year 2000. That impressive scale of growth would require a need for more structure to maintain compliance and consistency throughout the company.

This explosive growth wasn't exclusive to Koch, and it was the beginning of a shift in the general direction of corporate business in our country. The change would lead to more system-driven approaches and away from personal reliance. The company saw the need to have a common structure and formed Koch Operations Group—KOG. The leader of KOG was Bill Caffey, who, late in his career, would be the suit over the ranches in the early 2000s.

Bill was a good leader, but many of the new expectations were uncomfortable and sometimes led to becoming cynical. During a meeting of around 400 mid-level management employees, he gave us some wisdom that has influenced many decisions throughout my life. "Don't be a victim," he admonished. "We are only that by choice." Production, efficiencies, and profit would take a back seat to "compliance-to-regulation" for a business to be sustainable. Regulation compliance is a challenge that remains to this day. To their credit, Koch constantly preached, and practiced, 100 percent compliance, 100 percent of the time—at least in the areas where I was familiar. I appreciated that the ranch was a leader in doing things right and environmentally sound.

Right after Labor Day weekend in 1990, Marion and Jackie drove up to the Staudaher to visit with Sue and me. We were surprised to see them and welcomed them into our trailer-house. I could tell by looking at them that they were both pretty troubled and upset. The suits had just flown in on a company jet to have a meeting with Marion and to make him an "offer" he couldn't refuse. At least the suits treated him with respect, and they did take care of him. Marion wanted to let me know they'd soon be hiring someone to replace him and that both Steve Ingram, the farm foreman, and I would likely be considered. I was not happy about his news either, since Marion had been my best champion and a great leader. Marion Cross summed up the excellent relationship between him and me one day while conversing with two of his brothers. "I consider Ray the son I never had," he told them.

Shortly after Marion's visit, in September of 1990, I was interviewed in Wichita for consideration of becoming the general ranch manager following Marion's retirement. I had a lot of thinking to do.

As I did each time I advanced in my career, I questioned my qualifications and if I had enough previous experience and knowledge to take on such an enormous responsibility and challenge. I loved the ranch and the country where it was situated. I was confident that we could improve the management of resources in a balanced way that would benefit everything from the range to the ranch, our employees, and our customers. I was excited about this new challenge—the once-in-a-lifetime opportunity to manage this massive historic ranch and nearly 500 square miles of prime Montana rangeland.

It may have looked like an enviable opportunity to the outside observer, which it undoubtedly was. However, there were strings attached that weighed like chains on me.

CHAPTER 11
Moving to the Big House

"If a man's got talent and guts to buck society, he's obviously above average. You want to hold on to him. You straighten him out and turn him into a plus value. Why throw him away? Do that enough and all you've got left are the sheep." Alfred Bester

Wichita wanted me as general manager, but they also wanted to dictate significant ranch-level changes to me from the corporate level without considering how their stipulations would affect almost every family on the ranch and the ranch itself. Nepotism was a difficult issue within the Koch family and was reflected in corporate offices and most of their businesses. The ranches, probably due to the earlier influence of Sterling Varner and the four Cross brothers from Kansas, including Marion, were not typically affected as long as family involvement was on a call-as-needed basis at minimum wage.

There had apparently been a perceived past problem or system policy change because the offer of the position came with the cold and emphatic dictate:

"Your family will not work for the company."

For years, families had quietly provided the glue that held the ranch support structure together in remote and forgotten locations. Aside from the five-year cycle of professional painting, the ranch depended on families to keep the physical facilities looking like a showplace by caring for the landscaping and endless white-paint trim. Until 1990, they accomplished those jobs using their personal equipment—including lawnmowers and vehicles. The ranch owned only two or three push mowers but had huge lawns.

A busy county road ran right through the middle of headquarters, and the public definitely noticed the upkeep. Employees who didn't live at headquarters year-round lived in the older trailer court that was out of sight across the creek.

Family members hosted company guests for meals at the cookhouse. They were maids and sometimes hunting guides for visiting fishing and hunting groups—and they were happy to do it. Those were enjoyable times that were of value to all.

Families assisted the cowboy crew and farm crew with numerous tasks, from brushing meadows and spraying weeds to working with the cowboy crew and taking care of odd jobs. Wives swamped out bunkhouses and cookhouses, filled in for cooks, produced thousands of dollars of winter grazing on upper meadows by keeping them irrigated, and the list goes on. Some of those jobs were compensated, but many were not. Families with an ownership mentality who called this great ranch home—treated it like home.

Hearing that ultimatum, I felt like I'd been sucker punched. "Do I even mention this to Sue and flat decline the offer, or what?" I silently prayed. I recognized this dictate as a very real "spit in the face" to all those family members who had contributed so much through the years. I believe that the negative direction of government, society, and corporations toward nuclear families has been a severe downfall for our country. Banning ranch families living on the ranch, from working on that ranch, was the type of thoughtless thinking that fed right into the downward spiral of today's society.

Sue and I talked through this decision, keeping in mind the balance we strived for—God first, family second, and work third. She cried a little, feeling the callous sucker punch too. However, being an employee never really mattered to her, as opposed to being a ranch wife working with her family in a wholesome, engaged lifestyle that we loved. That's just how she's wired.

Once Sue figured out that they couldn't stop her from being my wife, my "help-meet," she let it go and fully supported me, knowing how much the ranch meant to our entire family. We both believed we were right where God wanted us to be.

"If I cared about being an employee, I would never have quit my good tech job," she assured me. "The kids and I can still help. The benefits of bringing the kids up this way are worth it. Besides..." she half-joked, "it'll be easier now to be a fair-weather cowgirl."

I didn't say it, but I knew it would never be easy for Sue to be "fair-weather." Her best photos, which I call "cowboy suffering photos," came from the most vicious days weather-wise, and she wasn't about to give up her passion for capturing ranch life with her ever-present camera—nor her passion for being a living, breathing part of the ranch itself.

We recognized that God had a purpose for us and had blessed us in our balancing of life so far. We accepted the challenge and never regretted it. It allowed us to raise our kids in a unique and wholesome way, improve the opportunities for other families, and help the ranch and the land to thrive.

We moved to the big old white house at ranch headquarters. There were already three ranch phones on the walls that rang in the kitchen, the bathroom, and the master bedroom of the manager's house. The sole private line was upstairs. The first time Sue mowed "our" gigantic road-front lawn with the ranch's self-propelled John Deere push mower, it took her nine hours of brisk pushing to finish in one day. She came in that evening with blisters on her hands and a determination that we needed to do something different with all that lawn. She began to fully appreciate the extent of behind-the-scenes work that Jackie Cross had faithfully and willingly performed for so many years.

Right off the bat I was hit with a perfect example of what to expect from the current Wichita suit who was calling the shots now. Some who had been in that position were mentors, advocates, advisers, and enablers to help the ranch managers. A few, like the current one, used that position to flex their muscles and pound their chest. I had total empathy for ranch horses who had to endure the same problems from constantly changing cowboys.

As the new general manager, one of the expectations I'd brought up to the company was the need for improved housing for our employees, beginning with the cow camps. I sat down with the suit in Wichita headquarters and Bob Kilmer, who at the time, was manager over all three ranches. Bob, like Larry Angell, had started with the company during the same period I started. Bob, his wife, Wanda, and their baby girl had lived on the Beaverhead Ranch several years earlier, then he'd been manager of the Texas Matador for several years. He knew firsthand the housing situation on the ranches—and the current structure in Wichita.

After a brief discussion, the suit told Bob and I to pick out the prefabricated housing we wanted to set up at the Jake Cow Camp for the crew, and to get back to him within a week. Bob and I went to work on that housing project along with all our other duties. Before the week was up, Bob called me.

"I hate to tell you this, but we have another problem. I've just been told to forget new housing. He wants you to figure out how to eliminate all employee housing on the ranch."

We did not get rid of employee housing, but the very opposite switch of commands shows how quickly and dramatically things can change while working for a corporation.

I had not had much involvement with the farm-side of the ranch when I became general manager. A considerable part of the overall ranch operation involved approximately 3,000 acres of hay ground around the headquarters area. In 1975, my first year, most irrigation was still accomplished with flood irrigation—ditches and dams.

There were some handlines, which were pipes with sprinkler heads that had to be carried one at a time to the next section of the field, or "land." The lines would then be reattached into a continuous pipe and hooked into the mainline—a very labor-intensive way to irrigate large acreage. I don't recall that there were any wheel lines at the time, but even Sheep Creek, which is irrigated by pivots now, was entirely under flood irrigation. We pumped water out of some of the deep wells to flood irrigate the gravelly Sheep Creek ground, which was only marginally successful.

Because the ranch put up all the hay in small square bales, a large farm crew was required—usually more employees than the cowboy crew. Flood irrigation demanded numerous seasonal irrigators. As we converted to handline and wheel line sprinkler irrigation, it took an army of workers, usually high school or college kids, to move that pipe. I recall a couple of college kids who moved pipe in the late 1980s. Each of them earned $9,000 that summer, moving pipe for two and a half months before going back to school. That was a lot of money for the 1980s and made them the highest-paid employees on the ranch, except maybe the general manager.

George Munski was the farm foreman for many years and did most of the flood irrigating in the Upper Blacktail meadows—a job that Rita Sawyer took on after George retired. When irrigating the 3,000 acres of farm ground in the summer months, there was usually an irrigator boss under George. He was responsible for going to town and finding workers—usually a crew of teenagers—for moving the pipe two times a day. It was a big job and challenging to keep a regular crew. Consequently, some of those handlines were converted to wheel lines. Wheel lines are manually moved, but an engine in the center of the line drives the wheels, moving an entire line. Until labor laws became too difficult, we resorted to hiring folks from across the southern border for a few years because we could no longer find high school kids or anyone else to irrigate.

The ranch then converted most of the old handline irrigation to wheel lines and eventually started upgrading to automatic pivots. Marion was still the manager, and George was the farm foreman when they started putting in the first pivots. Pivots weren't as efficient with water but converting to pivots was a good decision since the labor situation got to the point where we couldn't even find enough people to move the gas-powered wheel lines. We continued to convert our irrigation systems until the entire farm was under pivot. By 2011, we had seventeen pivot installations.

Pivot irrigation was well suited to the fields with gravelly soils. The deeper soils along Blacktail Creek on the east side of the farm held moisture well and were great candidates for flood irrigation. For reliable irrigation, though, automatic pivots were the future. Changing the way we irrigated also decreased the number of personnel we needed.

In the 1980s, the farm put a lot of acreage into grain. The year 1988 was the last year we grew malt barley—then still had to buy hay to feed our cows. Following a thorough analysis of the entire ranch operation by segment, it was apparent that farming and raising grain was not making us any money. On the contrary, it was costing the company, so grain farming was discontinued, and the farm concentrated on raising quality hay—mostly alfalfa.

The ranch owned an entire fleet of balers and swathers necessary for putting up all the hay in small square bales, the haying method we used until 1990. We hired out the hay stacking to independent stackers, primarily to Todd Stoddard's stacking company. After studying haying systems in 1990, the ranch switched to a round bale system. We researched all kinds of hay operations to determine which would be the most efficient for farming and feeding on the Beaverhead. We compared small square bales, round bales, big squares, loaves, and loose hay. Using round bales was the clear leader as far as labor for putting the hay up and feeding it out in the winter, so, for a few years, that was the route we took.

Later in the 1990s, when I was the manager, I had time to analyze ranch systems and see where the highest costs were tied up. I determined that we could keep a minimal farm crew responsible for managing the irrigation, getting the hay raised, and feeding that hay out, plus essential maintenance and repair. Everything else, the swathing, baling, stacking, and any farming that had to be done, such as ground tillage to rotate those crops, could be contracted out the same way we were already contracting out the stacking and annual feedlot cleaning.

In 1995, we had a big farm auction and sold all our machinery. We had a fleet of every type of haying equipment that sat idle most of the year. We sold it all, keeping only one tractor and a skid-steer. Taking care of the feedlot required a second tractor which we would lease. That was one of the best management decisions I ever made. We got out from under a lot of metal, the never-ending costs of maintaining that metal, and the not-so-small fortune we had tied up in equipment that was only used a few months out of the year. Selling all that machinery made us a lot of money. Even more importantly, we realized enormous savings by contracting the work out, not only in machinery maintenance and fuel but also in labor since we no longer needed to hire such a large farm crew.

The same year, we also switched the way we operated on the cow-side and our complete management strategy. We went from an average of twenty-eight full-time people, plus many seasonal people on the farm and cowboy crews, to fourteen full-time people, including me. After about five years, the company changed its stance on family. Now if we needed temporary help, we could put family sixteen and older on the payroll. Family members could be part-time help at minimum wage with a cap on hours and would now be covered by workman's compensation. That's how we continued until I left in 2011. Few people had any idea how hard and how long I had to fight to make this happen.

One of the things that's hard to imagine now is putting up 10,000 tons of small square bales, but that was the ranch tradition until the 1990s. It also took an army of men to feed all those small bales in the winter. We usually had about an eight-man crew split into a couple of different crews with hay wagons that would hand-feed small square bales to everything that needed feeding all morning long, and then they'd spend all afternoon loading back for the next day.

George Munski was the farm foreman for a long time and one of the longest-term employees that the ranch ever had. He started on the cowboy crew around 1968 or 1969 and retired after about thirty or thirty-one years. George could handle it. A great guy who could do anything and was the go-to guy by default, whether it was construction, irrigation, farm machinery, or home handyman. I'm sure there were times he would have liked to have changed his name to something besides, "Hey, George!"

He was a tremendous asset to the Beaverhead Ranch and kept us running. George knew where every water line at headquarters was, how to fix issues at the trailers with heating, plumbing, and sewer lines, and he always winterized the Staudaher Cow Camp and got it going again in the spring. He and Dick Everitt performed a lot of the minor remodeling in the old homes. His experience was invaluable to us through the years.

The highlight of George's career seems simple and can only be understood and appreciated by someone who has baled a lot of hay. His claim for greatest achievement happened one summer when he used only one shear bolt to bale 10,000 small square bales with a Freeman self-propelled baler. That is indeed a fantastic accomplishment, but George had a raft of achievements he humbly believed were all part of a day's work.

When the ranch began contracting out the actual equipment-based farming in 1995, the skill set and responsibilities of the farm foreman changed. The ranch's primary requirements were unconventional now. They needed a candidate whose experience and job requirements were pivot irrigation, feeding, and maintenance. Thus, finding a good farm foreman became more difficult, and the challenge continued.

Key people who had a long-term influence on the farm besides George Munski were Dick Everitt and Steve Ingram, who both served as farm foremen and other duties for some time. Dick's wife, Wanda, worked part-time in the ranch office.

1990 - Roy Cornell, DVFD firechief receiving donation from company. Marion Cross, Steve Ingram, Wanda Everitt, and Ray Marxer.
Susan Marxer Photo

Having achieved a considerable reduction in calving issues, we implemented a significant change from historical calving traditions. We reduced the crew responsible for calving and supplement feeding of 4,000 cows from six crew members to two. We would no longer use the primitive Jake Camp or hire a cook. The men could stay in their headquarters housing, and single men would eat at the main cookhouse. Thanks to some good employees and a willingness to change out of necessity, the change was effective.

When I started at the Matador, the business was structured very similarly to most companies in the mid-1970s, and the way many in agriculture still are. It was a traditional command-based system.

Koch had more organization and structure as a corporation, but the main difference was scale. The top-tier was the corporate board of directors and a president of Koch Industries, Inc. Under that tier was the president of Matador Cattle Company, who worked in the Wichita office. He was over the next level—the general managers of all the Koch ranches. At one time, there were five: Spring Creek in Kansas, Matador Cattle Company in Matador, Texas, the Beaverhead Ranch in Montana, the Roberts Ranch, also known as the Bug Ranch, in Wyoming, and the Garvey Ranch near Winnemucca, Nevada, which was a five-year lease.

Max Sweeney in Wichita was the controller—the accountant for all the company ranches. Max held that position for a very long time. He understood the nuts and bolts of the accounting system and how each ranch was doing, better than anybody else in the company. Each ranch had a secretary, who was also a necessary bookkeeper, so we could write checks and do payroll at each ranch. Around 1988, the payroll and bill paying moved to Wichita. We still had petty cash and a local bank account with check privileges, but the company switched the managers to using company credit cards for most purchasing.

At the ranch level, the top-tier was the general manager, who was the middle man between the Koch tiers, specifically the president of Matador Cattle Company and the people on the ranch. On the Beaverhead Ranch, the general manager was over the office employee and the different foremen, including the cow foreman, the farm foreman, and the Sage Creek foreman. Both the cow boss and the farm boss had large crews that they managed, and the Sage Creek foreman usually had two men and sometimes a couple of seasonal irrigators.

The structure was a command-based system where the decisions were made higher up and then filtered down to the people who would implement them. The company didn't share financial information, business information, and the like with anybody below the general manager level. The manager had access to the financial statements because he was responsible for filling out a budget each year, but the foremen never got to see a financial statement. The foreman's primary duties and tasks were to get things done. The tiers above us decided the tasks and jobs we were responsible for doing. That structure is still how many smaller and family operations operate to this day.

The lack of incentive or way to keep lower-level employees engaged and contributing beyond physical labor was one of the things I saw as a significant obstacle to ending the cycle of constant turnover, and it was an opportunity for improvement. Several others in the company and ranch management also recognized the need for change and began to take action.

In about 1988, the company started sharing more information with the foremen and involving us more in planning how we would operate and what we would do. They shared financial and industry knowledge and other helpful information, which gave us a better understanding of the business, which was essential if we would build a team that could progress.

We needed a team-building system rather than depending on individuals to carry the bulk of responsibilities. We were trying to build some institutional knowledge, and the way the command-based system was, nobody stayed long enough to do that. There will always be a beginning or lower tier, and some are content to remain there or leave. Others, however, thrive on challenge and responsibility. If they can't grow where they are at, they aren't likely to stick around—a lost opportunity for a company if that employee has the character, values, and beliefs that fit its mission.

Gary Mills had been at the Sage Creek ranch "training ground" like Marion and me. He became our cow foreman for the next few years. Due to our experiences in the selection and development of cowboys, we both recognized a need for change.

I wanted the ranch and individuals both to be successful, but realized that the employees and their families had to feel successful for that to be sustainable. We sat down to figure out the difference between long-term past employees and those who moved on after a short time. Some differed in fundamental values and beliefs, but most of the time, the difference was in decision-making and family involvement opportunities.

Those of us that had been at Sage Creek had to treat it like it was ours because if we had a problem, we had to fix it ourselves. Resolving issues created employees with an owner's attitude, not just a paid hireling. That would be the breakthrough and core of our strategy for change. Giving more employees "ownership" opportunities would be change created out of need, not desire. I credit Gary for having guts enough to help me design a new structure for management that would do away with the traditional cow boss job we both had experienced, and he currently held.

We created this new structure out of genuine need while Charles Koch led a company-wide change in management style called Market Based Management, or MBM. We did not know all the correct buzz words or acronyms, but we put into practice what would later become the norm for Koch Industries. Interestingly, some years later, Charles would mention our changes at the ranch in an address to the Austrian Parliament and also in his book, *The Science of Success*, the inference being that if cowboys can do this, anyone should be able to.

Some skeptics of our proposed changes said it would not work because we did not have the right people in place to make it work. My response was, "If we wait until we have the right people, it will never happen." We launched off into MBM and had some setbacks for a couple of years, but it changed the ranch and people's lives once in practice. This management-style change was probably the most significant of all the changes we made over the years. It addressed all the critical elements necessary for a sustainable business.

The former structure with a cow foreman and a crew established that the cow foreman had the role of an owner over the entire ranch. Thus, he directed the crew to perform tasks, generally concentrating on one portion of the ranch at a time.

This command-based structure left the large remainder of the ranch without a daily "owner." Our new system would require the crew to help each other as needed, but every portion of the ranch had an "owner" taking care of it every day. Having a specific place of responsibility was another management breakthrough that resulted in stable, productive employees who took on an owner's attitude.

CHAPTER 12

Koch Beef Venture

"Strategic vision is the ability to look ahead and peripheral vision is the ability to look around, and both are important."
Carly Fiorina

The 1990s brought a substantial shift in focus for Koch toward agriculture. The company grew exponentially with a vision to fully integrate our beef business. Koch Ag and Koch Beef were added as new subsidiaries of Koch Industries, Inc., and the ranches were placed under Koch Beef, taking on the new name—at least at the business and signage level.

The company's cattle enterprise went from the three existing ranches to including six large feedlot operations and a retail market plan for beef produced. Other key elements were a supply or cattle procurement division, grain division, and alliances with packers to get our product to the consumer. The vision, "Conception to Consumption," was the focus.

With the intense expansion came a significant increase in management positions at the corporate level. Koch was unable to fill those positions from within as the company had minimal experience in those areas. Koch wanted to do this at an immense scale and very rapidly. Numerous experts in the different fields were hired and put in charge, including analysts, traders, and nutritionists.

Rapid growth created a demand for corporate services that had previously only supplied the three ranches. In cowboy terms, we were already dealing with the recurring three-to-five-year change in "suit," compounded with the periodic five-to-eight-year change in the next level up, which I will refer to as "a wardrobe change." The sizable addition of Koch Ag and Koch Beef was like adding a multi-room wing to the house—that still only has one bathroom.

This over-extension of internal services created a challenge for ranch managers accustomed to receiving and using monthly financial reports to make decisions. Suddenly the ranches were no longer viewed as a priority and became the stepchild. We operated for the next two and a half years with only a year-end financial report in which we had questionable confidence.

A needed change came about on the Beaverhead when we created our own enterprise accounting spreadsheets, along with spreadsheets to manage livestock and feed inventories. The best ranch management education I ever received was the week I spent at *Ranching for Profit School*, led at that time by the founder, Stan Parsons. The enterprise accounting system and spreadsheet development we learned were invaluable in helping to understand our business and to make wise decisions accordingly. This financial recording system could greatly help most ag operations, many of which do not understand the costs of their individual enterprises.

The standard financial statement or annual balance sheet most ag operators receive from an accountant at tax time provides very little beneficial or timely information for making operational decisions. Many people fall into the rut of making numerous short-term decisions instead of long-term strategies. A principle question espoused in the Ranching for Profit School is one every ranch operator should consider: "Am I working in the business or on the business?"

We were not just guessing about the ranch status of priority because, on more than one occasion, the new "suit" over the ranches jested, "Why don't we sell these big SOBs and use the proceeds for something that makes us money?" Not exactly encouraging to hear from your leader and certainly not within earshot of your dedicated employees. One of my key employees, Steve Stafford, was there and listened to the crass remark. It was the last straw and caused him to leave. That was the loss of great potential for Beaverhead Ranch as that individual had the unique combination of values and beliefs and intelligence and skills so rarely found.

Steve helped us resolve recurring health problems in some of our weaned calves that we had been unable to solve despite lots of testing. He arranged a new research project with Montana State University to do liver biopsies on live cattle to determine what minerals were stored and available in the animals. The findings were that the animals summered in areas with dense conifer stands that contained high levels of molybdenum which made the copper and other micronutrients unavailable to the animal's immune system. An injection of BoSe at branding solved the problem.

Steve would use his expertise in animal nutrition and commitment to work for another large corporation for over twenty-five years. He was my choice to replace me then, and now, some thirty years later, I still believe he should be managing Montana's Matador Cattle Company—regardless of who owns it. Even after all this time away, Steve still has that care and love for the land called the Matador that few ever will.

Steve Stafford

While Koch Beef did not consider the ranches of much consequence, we continued to improve. We made the most significant changes at the ranch level during this time despite not being considered essential to the big picture of Koch Beef. The slight was magnified even more with Koch's massive purchase of Purina, the addition of yet another wing on the rambling, multi-level "house with only one bathroom."

During my first decade of management in the 1990s, the Beaverhead Ranch made phenomenal improvements across every system:

- We restructured our management at Montana's Matador, and improved our selection and development of employees, cows, and horses.

- We gained a better understanding of our products produced, and improved them through a carefully designed crossbreeding system and artificial insemination.

- We increased weaning weights by 140 pounds while at the same time, we reduced feed costs by 33% and increased ranch carrying capacity by 8%.

- We shortened our breeding season to forty-five days. Remarkably, in our entire herd of about 6,500 mother cows and heifers, 80% of all calves were born in the first twenty-one days.

- We reduced full-time employees from twenty-eight down to fourteen while improving employee retention from an average of two years to eleven. We shortened our workweek to five and a half days for most of the year.

A crucial change that would allow us to do more with less was taking advantage of goose-neck stock trailers. This piece of iron has actuated a paradigm shift in cattle ranching. When we replaced the cow foreman position with four herd managers responsible for different ranch divisions, we provided each herd manager with a sixteen-foot stock trailer and a ranch pickup. We even furnished a truck and trailer to one of the support help in the Centennial Valley. The flexibility of scheduling allowed the cow crew to work independently on their own divisions, team up with one of the others, or work as a crew for substantial jobs such as branding and weaning.

The massive shift in how we approached production across the entire ranch significantly increased labor efficiency while reducing the number of laborers needed. The change in labor reduced or eliminated the need for bunkhouse and cookhouse facilities. The only downside was the reduced miles required of our horses—although they still got plenty. Fewer challenging, extended rides reduced the quality and toughness of our horses. As cow boss, when it became too difficult to find applicants who could shoe horses, I had already begun hiring professional farriers to shoe ranch horses. Professionals saved time, injuries, and poorly shod or lamed horses.

The machinery auction in 1995 that eliminated most ranch equipment also contributed to our streamlined operation. I've already credited the most influential teacher for the principles we applied to achieve these profitable and productive changes. Stan Parsons, one of the most effective instructors I was privileged to learn from, no longer teaches the Ranching for Profit School. The school continued however, owned by his protégé, Dave Pratt until 2019. New owner, a former associate and leader, is Dallas Mount. One foundational principle I gleaned from the school, was that beef production is a separate enterprise from landownership. The division of enterprises is a critical concept for a profitable cattle business. Most ranches maintained long-term stable ownership because land continued to appreciate at a high rate while operational returns averaged 2% to 5%.

Land stewardship and range productivity on Montana's Matador also began a noticeable uptrend and improvement during the time when the ranches were considered more as hunting playgrounds than businesses by some of the "suits." Consequently, these massive paradigm shifts on the Beaverhead Ranch, accompanied by increased profits, went unnoticed and unappreciated. I received no raises in compensation during this five-year time frame.

Maybe it was because I had turned down the company's offer to manage all three ranches—an offer that was out of the question since I would not move my family to Wichita or work long-distance. At first, I thought my refusal had eliminated any bonus. But after five of the most profitable years the Beaverhead Ranch had ever known, I figured all our accomplishments had been overlooked as other enterprises were deemed more important. Ironically, our own parent company was unaware of most of these changes and achievements until Montana's Matador Cattle Company began being recognized for environmental stewardship, and our accomplishments were mentioned in many news stories, videos, and publications.

It was also ironic that the Beaverhead Ranch's environmental stewardship examples would ultimately become of more importance than the dwindling Koch Beef experiment and Purina venture. Both of these ventures would experience what I would call a "house fire" with only the primary residence remaining. It was interesting to see that the remnant of the whole Koch Beef venture were the three ranches that started it all.

I am disappointed that Koch Beef was not successful with the beef vision as it addressed many of the inefficiencies of the whole beef "Conception to Consumption" system. We gained a great deal of knowledge that would benefit us in the future. One of those genuine benefits was gaining an understanding of the beef industry from the big picture view that is lacking at the cow-calf level. At a CattleFax Cattlemen's College, I was shocked to learn that 50 percent of the nation's cow herd was east of the Mississippi River in herds of twenty or less.

Understanding the role that each industry segment plays is beneficial in creating and maintaining a profitable and sustainable business. I recognized that the packing and feeding sectors knew what drove their profit and efficiency much better than the cow-calf segment. Not understanding the feeding and packing segments was an obvious negative for the cow-calf sector because our customers knew what drove profit for their segments better than we did. Thus, the cow-calf sector continued to be price takers, and the packers and feeders directed the changes in beef cattle in general, becoming the price makers.

The feedlot guys wanted larger healthy cattle that they could run more feed through and receive higher gains. The packer wanted larger cattle with lower yield grades and higher quality grades. Neither the packer nor the feeder was ready or willing to pay for those wants when they were made available. Certified Angus Beef (CAB) was a good example. Some of those wants are antagonistic to what drives profit at the cow-calf level, so it required balancing traits for importance as the market continued to change.

At the ranch level, a breeding decision change takes a minimum of three years to see the result of that change. Realistically, the financial incentive may have changed in the three years it took to create the want. This scenario continues to be typical today, mainly because the cow-calf level does not know what drives their profit as they should.

From conception to consumption, most of the cost and value of beef is created at the cow-calf level. The improvements in value creation and the capture of that value at the packer and feeder phases are incremental. The feedlot can only add weight, and the packer can only provide harvest and marketing efficiencies to an already designed animal.

The most significant opportunities for system improvement in efficiency and profit reside at the cow-calf level, where product design and development occur. Those opportunities will continue because of the unique ability of a cow to transform forage grown on the earth's large percentage of non-tillable land into something humans can eat. Hogs and chickens are much more efficient at converting grains into something for human consumption.

I cannot emphasize enough the importance of having a cow that fits her environment. It is not the typical 1,300-pound or larger cow.

In 2007, 2008, and 2009, the average weight of all cows sold at the Beaverhead Ranch annually, which was approximately 1,100 head of dry, open, and cull cows, was 1,107 pounds.

At the ranch level, the highest cost is non-grazeable feed. That is where the most significant opportunity lies as well. A cow that can graze year-long with few supplemental feed costs will be much more profitable than those that require alternate feedstuffs. We had already confirmed that fact from having made the switch in 1970 from haying meadows to reserving them for winter grazing for cows.

For example, in 2007, 2008, and 2009, our supplemental feed cost was $84 per cow, including the cost of hay raised and fed. Compared with like-climate ranches, most northern region operations had costs that exceeded $300 per cow.

When I became manager of Montana's Matador in 1990, we had good grazing plans and established carrying capacities on our leased lands. However, it was experience over the years that established grazing plans and carrying capacities on most of the private lands. Knowing what those numbers were—how many cattle we could run in a pasture for how long—required institutional knowledge expected of the ranch manager. The ranch did not have a written record of pasture size or capacity for anyone to use for decision-making. Since I'd developed an analytical approach to such things to make the best decisions, I created another spreadsheet.

This grazing spreadsheet included the name of the pasture, size in total acres, and different ownership acres—private, state, or BLM. Then I used an MSU Extension service guide to help me establish carrying capacity. It was called *A Guide for Planning, Analyzing and Balancing Forage Supplies with Livestock Demand*. This simple guide used the land's topography described as "run in, run off, or normal" and general observations of range conditions to establish a baseline capacity for each pasture.

I also included an "available season of use" since some lands are only grazeable in the summer months. I included the tillable land or farmland in this as well. The spreadsheet converted the grazing capacity to available AUMs—animal units per month—even on the farm winter-grazed hay acres, using three AUMs per ton of hay. I inserted a rainfall variable to help with annual changes as well. The spreadsheet calculations allowed me to establish what AUMs were available each month.

Then, I made a second spreadsheet with all the different classes of livestock on the ranch and their monthly feed needs that would combine with the available AUMs to develop a grazing plan for the entire operation systematically. The herd managers could then use the results to make calculated decisions in the future.

I spent many hours with a planimeter and aerial photo maps of nearly 500 square miles to accomplish our Beaverhead Ranch grazing plan. As I recall, there were 148 separate pastures in this project in 1991. One future benefit of this project was that we had a spreadsheet to quickly help us create a feed budget for the coming year by projecting what the feed needs would be, and what would need to be supplemented to the grazing to meet those needs.

As with all similar projects I tackled in this manner, I gained a more thorough knowledge of this ranch called the Matador. As I look back I realize how important this grazing plan, had it been continued after I was gone, would have been when more land acquisitions were added to the ranch. The plan would have shown that most of the additions were summer grazing areas that only provided seasonal AUMs, throwing the balance of year-long grazing and livestock demands (cow numbers) out of whack.

Consequently, an increase in cow numbers would cause a need for costly alternate feed, thus reducing profit. The increase in numbers could work if it were in stocker-grazing for the summer months. The other profitable alternative would have been to purchase (if available) more acres in vegetative production similar to the Blacktail meadows if the objective was cow herd expansion.

I enjoy innovation in ranching, where perceived obstacles become possible opportunities. The Upper Blacktail meadows were the heart of the ranch since 1865, when ranch founders Poindexter and Orr had to leave their cattle behind when the onset of winter prevented them from trailing a herd over the mountains. When the two men returned from California in the spring, they found the cattle they'd left behind in the Upper Blacktail, fat and flourishing. Those cattle had spent the winter in the same meadows that the Matador had been winter grazing since 1970. I wanted to figure out how to mimic the winter and spring grazing on those meadows to accomplish similar results on the farm acres we had in alfalfa hay production.

One obvious obstacle was that alfalfa left standing dries up and loses most of its nutritional value as the leaves fall off, leaving only tough stems. If we could capture the second cutting volume and quality without baling, stacking, and feeding, we could save a lot of expense. How could we accomplish this and store swathed alfalfa for later consumption?

I contemplated the storage idea, and how to avoid killing vegetation laying underneath swathed hay. "What if we swathed the alfalfa late, around the first frost, and left it in the windrow for storage?" I also questioned how windrows would stand up in the face of southwest Montana winters. "Moisture and wind might be a problem," I figured, "but if we combined two windrows into one large windrow, perhaps it would act similar to a loose haystack."

I decided to experiment first before asking for help from the crew. Windrow grazing was my idea for better efficiency, and I didn't know if it would work. I picked one of the smaller pivots, swathed it after the first frost, raked two windrows together, and left it lay. Waiting until the first frost allowed the grass to enter dormancy so windrows laying on top wouldn't cause damage to vegetation from lack of sunlight. At the same time, I installed a single-wire electric fence to experiment with controlled grazing on this windrowed pasture and in another area where we sometimes ran over 300 mature bulls through the winter.

Fortunately, both experiments were very successful. The hay stored well, and forage testing in January showed little to no loss of quality or nutrition. We did, however, lose the cost of baling, stacking, and feeding, which was a savings of at least twenty-seven dollars per ton. The cows fed themselves—even with a few inches of snow cover—and put all the manure back on the field where it had originated, spreading it evenly across the field at no cost. After the second day, all those bulls stayed behind that one electric wire. The crew was able to see how windrow grazing could work and reduce their workload. In years to come, they expanded the use of both the windrow grazing and temporary electric fence.

Change is always easier to accept when it's your own idea. Until the people carrying out new or innovative practices believe it can work—it won't. Numerous neighbors and interested parties thought I had lost my mind when they saw that beautiful second-cutting hay laying in windrows into and through the winter until we needed to graze them. The cows turned the windrows of hay into needed nutrition and a windrow of manure that would replace our need for fertilizer in the future. We regularly alternated where the windrows laid to distribute the manure evenly.

The electric fence would see use in many applications in the future. Once the crew saw how the windrows and electric fence removed the need for them to have to feed hay physically, they were happy to divide pivots with temporary electric fence.

At about the same time, we recognized the need to change how we grazed the Upper Blacktail meadows in the winter. We usually put about 1,400 to 1,600 cows in each of two pastures from December through March. Each pasture was about 850 acres. The traditional way we'd grazed since 1970 left the quality of the forage, and the quantity, diminishing as April calving approached. An increase in protein supplement was needed to maintain cow condition.

We used electric fence to divide each of those pastures into five smaller parcels and confined all 1,500 head of cows into one parcel at a time. Thermal springs provided water with no need to chop ice all winter. We would divide the time needed, and forage available per parcel, to get us through the entire period. Dividing the meadows up allowed for more even distribution of quantity and quality of forage, with less supplemental protein needed and better cow condition maintained.

Another obstacle that turned into an opportunity was the Blacktail bench. A couple of 5,000-acre bench pastures had limited water at each end, resulting in poor utilization of the centers of the pastures and limited available AUMs. Sue and I checked the water sources and the possibility of running water to central tanks. I wanted to experiment with grazing cells.

Our son, Clayton, helped me design and build these grazing cells as a 4-H project.

We installed gravity-flow water tanks and two electric fence grazing cells that each had nine pastures. One of the cells was mainly state land, so we had to get approval from the state. By rotating through the nine cells in a controlled manner, we increased available AUMs from 520 to over 1,400 while improving the range and providing more rest to the entire area. The increased productivity of the land provided an increase in revenue for the state and allowed our yearlings to stay on the bench longer.

Clayton was twelve years old when we started the two bench cells, and he earned the opportunity to manage the movement of the yearlings through both cells. He maintained the wire and insulators, checked the water, and learned a lot about forage manipulation and moving and handling yearling cattle. Having his own "job" (non-profit) was a big deal, and he carried it out faithfully. The project earned accolades from 4-H, and experience for future positions with other ranches, the US Forest Service, and NRCS.

Clayton was helping and learning much on the ranch during his teenage years. He started working for John Erb doing odd jobs and building new pens at the auction after getting his driver's license at sixteen. Clayton and Kristy both worked at the livestock auction on sale days. I encouraged him to work part-time for John Erb on his ranches if he got a chance. I knew John to be fair, but he also held high-performance expectations.

While working for John, Clayton did a lot of riding in the Blacktail and in Erb's Middle Fork and Antone country that John later sold to the Matador. In 2006, Todd Sawyer, who managed that land acquisition and the rest of the Blacktail, was unable to ride due to an injury, and his support help had quit. Since Clayton was familiar with the new country and was looking to return to Montana from Idaho, the company allowed Clayton to return to work for Todd under Kyle Hardin, the assistant manager, and live in one of the old double-wide trailers.

Clayton Marxer and Braxton - 3rd generation
Amber Burch with Asher

CHAPTER 13

Environmental Stewardship

"All I have seen teaches me to trust the Creator for all I have not seen." Ralph Waldo Emerson

2001 - Matador Cattle Company President, Tim Durkin, Charles & Liz Koch, Sue, Ray, Clayton, Anna and Kristy Marxer   *Susan Marxer photo*

Sue and I had joined the local stockgrowers association and the Montana Stock Growers Association (MSGA) on our own—and on our own dime—when I was still cow boss. We actively participated on various committees—including environmental, public lands, and endangered species committees. As manager, I represented Montana's Matador Cattle Company in the MSGA and the National

Cattleman's Beef Association (NCBA). Sue maintained her personal MSGA membership.

In 1998, the staff of the Montana Stock Growers encouraged us to apply for the annual Montana Environmental Stewardship Award. They were aware of many of our leading edge land stewardship and industry efforts, having attended several tours and production seminars we hosted at the ranch. Sue and I understood the importance of sharing the critical and positive role that ranchers play in maintaining open space, abundant wildlife and fisheries, and providing safe and healthy food and fiber for the consumer with an increasingly urbanized public. Putting a family face on agriculture and sharing our story in a positive way was an incredible connection with American consumers who appreciated our passion and heartfelt communication.

We got the Montana ESA application—a long, thorough, and daunting packet. Shortly after receiving the packet of pages, Sue and I were headed for Joplin, Montana, for a bull sale. Joplin was almost on the Canadian border—a long drive that kept us captive in the cab of that truck for several hours. Sue brought her briefcase with the lengthy application, and we went to work writing out the specific, detailed information while en route to the bull sale and continuing the process on the return trip. It was an eye-opening revelation of just how much we'd accomplished by simply doing what was right for the resources, and for the cattle that harvested the renewable grass resource. By the time we finished the application and the long list of actions from the standpoints of resources, business, and public communications, it was evident that those efforts contributed to an overarching goal of operating a ranch business that had sustainable profit and healthy land and water resources.

We approached various land and wildlife agencies which the ranch had worked with, for letters of support, and Sue chose supporting photographs from her extensive collection of ranch photos. Beth Almond, from MSGA, polished the application and prepared our final submission in one neat, beautiful presentation. That initial application, in 1998, was the first accounting of environmental stewardship on Montana's Matador Cattle Company ranch.

Knowing that a major corporate-owned ranch is not generally as appealing to the audience as a multi-generational family-owned ranch, our expectations were realistic. I addressed some critics by reminding them that effective land stewardship practices required human effort to learn and apply sound management principles regardless of the type of ownership. We were an example of a corporate employee, with a dedicated family, who was committed to the health of the business and the land. I considered it a bonus that we could be an example for others while positively influencing nearly 500 square miles of the West.

The rest is history. We won the 1998 MSGA Environmental Stewardship Award (ESA), making Montana's Matador eligible for NCBA's Region Five ESA. We would compete with award-winning ranches across the western states. In July of 1998, we were awarded the ESA for Region Five in Denver, Colorado. In January 1999, our entire family traveled to the NCBA Convention in Charlotte, North Carolina, for the final round of competition. Meeting with the other award winners from New York to South Dakota (our toughest competition), California, Iowa, and Florida was a treat. Too often, we never think of places like New York having ranches and ranch people.

An entire day was scheduled for media training. Regardless of who would win the national award, everyone there had a great story and needed to be prepared to share it. Not all media is friendly, and training was beneficial for preparing a clear, concise message that would be ready at a moment's notice. Being able to respond to media in a friendly and positive manner is a tool everyone should have in their toolbox.

The day before the award ceremony, our family was en route to our hotel, and Sue was quizzing the kids, ages seventeen, sixteen, and fourteen, on how they would respond to specific questions. She got frustrated with Clayton when he responded to a question about life on a ranch with a typical teenage tease. Dropping his voice deep and stealing a line from a Garth Brooks song, he chanted with the perfect beat, "It's boots and chaps. It's cowboy hats. It's spurs and latigo…" Sue grouched at him and gave it up.

The following evening, Montana's Matador Cattle Company was awarded the national NCBA 1999 Environmental Stewardship Award. After accepting the award, I had a few minutes to speak. The award was an honor I wanted to share with cattlemen in that room and across the country—ranchers who practiced many of the same principles in taking care of the land and the beef business. I mentioned the incident between Clayton and his mom. "You know, I thought about that," I said after the crowd laughed. "Clayton wasn't that far off. That's pretty much the life he's known."

At the NCBA Convention, every rancher is familiar with the little red IRM notebooks designed for shirt pockets, wherein are jotted down valuable notes for record keeping. I remember my parting thought when I said, "My favorite little red book is not the one I carry in my shirt pocket. It's the old-fashioned red hymn book in the back of our church pews—and the song that comes to mind is, 'To God be the Glory, great things He hath done.'"

Later, some award selection committee members told me that their final decision was difficult. The conclusive element that influenced their final decision was the commitment and involvement of our family. What an honor and a blessing that God not only allowed us to be stewards of that great resource known as the Matador Cattle Company through our employment with Koch Industries, Inc., but He also blessed our efforts above and beyond anything we could have ever imagined.

1994 - Trailing to Centennial Valley - "Notch" in background

In the late 1980s, Sue had a few freelance stories and photos published in *Country* and *Country Woman* magazines by Reiman Publications. One particular "Day in the Life" story had a tremendous response and led to interest from World Wide Country Tours (WWCT), an arm of the same Reiman company whose tour specialty was off-the-beaten-track bus tours which visited many country places featured in their publications. The Matador became a favorite stop on their

"Big Sky Country Tour." The tour originated in Salt Lake City, made a loop up through Idaho and southwest Montana, then swung down through Yellowstone National Park and Jackson, Wyoming, on their return route to Salt Lake City. We were step-on guides, meeting the coaches at Monida and eventually ending up in Dillon, where 4-H clubs hosted a pitchfork-fondue meal, and the Dillon Junior Fiddlers provided entertainment.

Our little corner of rural America with the vast ranch, friendly cowboys, and polite, active country kids never ceased to amaze the travelers. It made the Big Sky Country Tour and this particular stop perennial favorites. That public outreach for telling and showing the ranching story, and our beef promotion via the unique pitchfork fondue, played a part in almost every award the Beaverhead received.

Sidenote—pitchfork fondue involves cast-iron vats like the old-timers used for rendering lard, regular farm store pitchforks, and tender steaks threaded onto the tines. Joe, Jack and wife, Myrna, Champine would render beef tallow down to oil before plunging pitchforks stacked with steaks into the hot oil for four minutes to achieve the tastiest steak most of the travelers had ever experienced.

Sue and I and the kids continued as step-on guides for three to five tours per summer for over twenty years. When our newest "suit" learned in 2010 what we'd been doing—making ranch stops in the Centennial or up the Blacktail, and letting travelers visit and get pictures taken with genuine cowboys, he quickly stopped the tours.

Every year we reached 200 to 400 people, positively portraying the ranch, our employer Koch Industries, Inc., the beef industry, and natural resource use. Several seasonal employees we hired found us via exposure to the tours and the *Country* magazine articles and photos. We hired men from Kentucky, North Carolina, California, and a young man named Mike Blass from Ellisburg, Pennsylvania. Mike didn't have much experience but was a veteran, athletic and had heart and want-to. He was good help that branding season wrestling calves. Our kids enjoyed Mike and they spent a lot of time together. Mike is one that accepted Christ as his personal Savior before he left. A couple of years later, he showed up with his new bride, Laura, to visit and was excited about his new life and ministry. We stay in touch, and it's been such a joy to watch their happiness and family growth. We made many friends through those years of tours and would see some travelers a second time when they returned on the same tour.

We appreciate the opportunities and experiences gained through employment with Koch's Beaverhead Ranch that led to being national spokesmen for the beef industry. The beef association ESA awards were not the first recognition for land stewardship that the ranch received, nor the last. In 1992 the Bureau of Land Management awarded Montana's Matador Cattle Company the national Partners in the Public Spirit Award for our efforts and commitment to the *Sage Creek Rest-Rotation Grazing System Demonstration Area* that began in 1975.

1992 "Partners in the Public Spirit Award" presented to Montana's Matador Cattle Company. Ray Marxer, Ranch Manager, Gus Hormay, Rest-Rotation Founder, Marion Cross retired Ranch Manager. Photo Courtesy BLM Dillon Office

We were honored to share that award with August L. "Gus" Hormay, who developed rest-rotation grazing. I requested that Marion Cross, who had already retired two years earlier, be included in the presentation since he was responsible for the project's inception. The grazing system designed for the 80,000-acre Sage Creek Ranch would be the foundation for future grazing plans across the entire ranch. Rest-rotation would influence how we planned, executed, and measured land use, including photo-monitoring and grazing harvest data.

Influence which was gained through historical data collection and photo-monitoring, would be used in 1990 when a new federal land-grazing conflict arose. The newly contrived concern was that "riparian areas"—a blanket term for areas around water sources and sub-irrigated meadows—could not sustain livestock grazing. Ranchers knew those natural landscape features had survived large ungulate grazing for centuries. Opponents of livestock grazing on federal lands didn't care. "Riparian areas" became the new buzzword for an excuse to remove cattle from historical grazing leases. The issue arose in 1989, driven by a young USDA Forest Service range conservationist who majored in environmental science at a university back east.

Marion and I, and the Forest Service, had numerous frustrating tours and horseback rides trying to find a resolution. Marion Cross, one of the most patient men I've ever known, was still ranch manager at the time, and had reached the end of his patience. He proposed to the company that we legally fight the Forest Service decision requiring the removal of the 600-head herd after only two days of the normal twenty-one-day grazing period allotted. Marion was in the process of retiring, and I was assuming the role of ranch manager during this debacle.

Beaverhead Ranch had to determine a route for resolution. We felt a responsibility to federal land grazers across the entire West since the outcome of our fight could have a domino effect throughout the entire western cattle industry. Advice from Charles Koch, "the government has all our money they need to litigate this," helped us avoid a legal battle.

I aspired to discover a conflict solution and not perpetuate the conflict. Neither side had scientific data to validate their opinion. The USDA Forest Service had some records, but no data, while permittees had nothing but observations and memory. No available facts were conclusive enough to justify radical grazing practice changes. I obtained approval to seek advice from Resource Concepts, a resource advisory company in Nevada. John McLain, the principal, and associate Don Henderson studied the situation and concluded that we all needed more facts and gave us a couple of ideas of how we might proceed.

The proposal I suggested was, rather than continuing in conflict with the Forest Service, that we should invite them to partner on a five-year graduate study of the effects of grazing, along with other influences on riparian health in the contested allotment. Each partner would contribute $5,000 toward the study, which a Montana State University graduate student would conduct. The Forest Service agreed to the plan. For a grad student recommendation, I contacted Dr. Clayton B. Marlow, Professor of Range Sciences at Montana State University, who I knew to be a respected expert on Montana rangelands. Marko Manukian was the grad student selected to carry out the study. The ranch provided horses and summer living accommodations at the Jones place cabin in the Centennial.

While the grad study was underway, Sue and I established an extensive photo-monitoring plot for the allotment, which we monitored for the next twenty years. Our three young children accompanied us on horseback, making it a family affair. We used knowledge gained from Gus Hormay and BLM to establish the photo points. Our plot selections needed to represent what was going on in the whole allotment and not specific areas that could prove our opinion. It's a simple stratagem to find sites to prove a desired outcome, an often-used practice of litigious NGOs (non-government organizations), and some agency folks opposed to livestock grazing. We wanted to record what was happening over time, both in sensitive areas and in broader areas representing the entire pasture.

Clayton and Ray Marxer - USFS Lone Butte Pasture 8/25/90. Day 1 monitoring study. We had to move cattle because the USFS tech claimed the riparian zones were overused.

Marko diligently conducted detailed streambank disturbance and stability measurements and calculated grazing effects on willow colonies. His final analysis produced a thorough scientific master's thesis that allowed the ranch and the Forest Service to resolve the conflict, allowing the ranch's historical grazing to proceed with some changes in riparian management. That study had a significant impact well beyond Matador's single allotment.

The grad study findings provided timely scientific facts and data, supported by our monitoring photos, that helped to settle the 1995 "Beaverhead lawsuit." The National Wildlife Federation and their Montana affiliate brought that critical lawsuit against the Beaverhead-Deer Lodge National Forest in an attempt to stop livestock grazing using the year-old National Environmental Policy Act (NEPA) analysis regulation. Had that lawsuit succeeded, litigant groups intended to use the Beaverhead-Deer Lodge National Forest as a model for blocking grazing allotments across the West.

Grazing challenges on Federal lands are an ongoing battle in western states where the government holds a large percentage of the land which is typically interspersed with private lands.

The ranch continued to receive recognition for our land stewardship. We received the Region 8 EPA Regional Administrators Award for Cooperative Excellence in Denver, Colorado, followed by the first-ever International Association of Fish and Wildlife Agencies Private Lands Stewardship Award in San Francisco, California. This award was sponsored by the International Association of Fish and Wildlife Agencies, Wildlife Management Institute, the Wildlife Society, American Fisheries Society, and the American Farm Bureau Federation. This IAFWA award was significant since some sponsors were generally somewhat opposed to livestock grazing.

Koch Industries received the National Facility Award for Beaverhead Ranch environmental efforts from the National Association of Environmental Managers. In the spring of 2000, Sue and I were invited to Washington DC by the Department of Agriculture to join ten other individuals across the country who had done exemplary work in environmental management. Montana's Matador Cattle Company was one of the recipients of awards presented by the Deputy Secretary of Agriculture, Richard Rominger, at an Earth Day event held on the Capitol Mall in Washington DC.

Sue and I had gotten acquainted with Max Peterson, chief of the USDA Forest Service, and his wife, Jan, while in San Francisco for the IAFWA award. They graciously extended an invitation that we would stay in their Virginia home when we traveled to DC. Max was still the head of the USDA Forest Service and had served under several administrations. There was some meeting of world leaders in the capital, which required heavy security. Several streets were closed to traffic, as were many of the museums. Having a seasoned resident guide was an immense help. I had to leave a day earlier than Sue, and she used the extra day to take a personal tour of the Pentagon with Mrs. Peterson, who had worked there for many years. That trip, and the generous hospitality shown to us by the Petersons, was a special and unexpected perk.

Later that year, I was a finalist for the US Conservationist of the Year thanks to a nomination by the foremost champion of "America's Outback," C.J. Hadley of *Range magazine*. The Beaverhead Ranch also became the first ranch to receive certification from the Wildlife Habitat Council in Silver Springs, Maryland. The innovative management we put into practice in our cattle herd, on our land resources, and personnel management received a lot of good press. As a result, we received many speaking opportunities for various organizations from Montana to Arkansas, West Virginia, North Carolina, Washington DC, and Calgary, Canada.

Most of the time, our presentations were visual, using Sue's photos of life and work on Montana's Matador Cattle Company that illustrated the health and vigor of the natural resources, the cattle, and the people, including family. I usually had a good idea of the narrative to fit the need, a view of time, and arranged the slides as a guide for an off-the-cuff presentation sensitive to the questions and interests of the audience.

It was impossible to talk about the ranch and our family without recognizing that none of this would have happened without God. He remains in control regardless of man's efforts to wrest that control. One thing that riding endless miles on horseback over the high, wide, and lonesome, rocky sagebrush hills does, is remind us how small and insignificant we as humans really are. Our ultimate purpose for this book is not to bring accolades or fame to ourselves, but rather to witness what Jesus Christ has done in our lives and through our lives and what He can do for you if you would only receive Him while there is still time.

"Boast not thyself of tomorrow; for thou knowest not what a day may bring forth." Proverbs 27:1

CHAPTER 14

Roles, Responsibilities, and Expectations

"Corporate culture matters. How management chooses to treat its people impacts everything—for better or for worse."
Simon Sinek

Beaverhead Ranch continued to effect change and innovation into the first decade of the twenty-first century, but at a slower pace than in the 1990s. We all learned a hard lesson from the Koch Beef venture and the Purina venture—"Learn to walk before you run." Profitable production and meeting customer needs became the dominant focus for Matador Cattle Company. In general, the years 2000-2009 would be a time of fine-tuning and growing the many ranch business innovations that had transitioned our traditional structure into a more streamlined and employee-friendly operation.

The breakthrough changes that highlighted the 1990s would continue to influence ranch growth in smaller increments and improvements. We dealt with numerous "suits" and a few ranch personnel changes. Kyle Hardin, who came to the Beaverhead Ranch in 1988, fresh out of Texas Tech, and worked for about five years, returned in 2001, bringing talents and technical capabilities we would need in the future. Kyle didn't use his degree much the first time since he moved irrigation pipe and later worked on the cowboy crew, but as manager, I needed his administrative and technical talents.

Kyle was a people person and a big help to me in preparing reports for the Wichita office, managing compliance, coordinating the ever-increasing ranch meetings, and helping with branding, shipping, and cowboy support for Todd Sawyer on the Blacktail side of our summer range. He was good natured and always ready and willing to contribute wherever needed.

Following the formation of Koch Beef and the addition of the feedlots, Koch Beef needed management-capable employees at the feed yards. While we would miss Kyle's support and pleasant attitude, I felt good about helping someone advance beyond the ranch to grow and enhance their talents. Kyle managed a couple of the feed yards until Koch Beef liquidated. He then went to Purina until Koch divested of that as well. I had lost Steve Stafford, my assistant manager, and was pleased Kyle would return to take on the assistant manager role. Kyle became ranch manager when I left in 2011.

The ever-increasing government compliance issues demanded more Herculean effort, time, and expense than in the past. The original feedlot at headquarters was built in 1948 before Koch purchased the ranch. Constructed with wooden poles and feed bunks for feeding chopped hay, the feedlot was severely in need of replacement. The ranch used the twenty large pens primarily for replacement heifers and weaned calves. We undertook the major project of replacing and rebuilding the feedlot and the necessary design of new drainage systems.

New regulations on the horizon prompted the ranch to consider a CAFO (concentrated animal feeding operation) permit from the DEQ (Department of Environmental Quality). A permit may not have been required at the time, but we wanted to be ahead of future determinations.

Koch was very proactive on compliance issues. With their support, we went beyond regulatory requirements to rebuild the feedlot according to CAFO regulations and permit requirements. The Montana DEQ was a great help, appreciating our proactive approach. We dismantled the old wood pens and reshaped the ground to accommodate an engineered drainage system and collection pond that would capture all runoff from the pens. The original setup was designed to prevent water from running into the pens, which was a plus.

We built the pens out of pipe with continuous concrete bunks, which were probably the first in Montana. We had already installed continuous-flow water tanks from a gravity flow thermal spring several years earlier. While we were at it, we also modernized other corral and handling facilities. The CAFO permit would require numerous new monitoring efforts and nutrient management plans to be compliant. Kyle Hardin would use his feedlot experience to help us navigate the additional compliance issues.

During the first decade of the twenty-first century—the 2000s, we made employee housing improvements for families a priority. We replaced the old single-wide trailers in the two trailer parks with a couple of appealing modular homes in each location. To improve quality of life, we strategically selected and developed a few new housing sites that would provide more scattered and private locations. We had long-tenured singles, but the majority of stable employees had families. If families weren't comfortable in their housing and ranch location, the tenure of the employee would likely be short.

Dramatic changes we made to our breeding herd genetics improved production efficiencies substantially. The use of proven sires in our AI program on replacement heifers improved the calf-crop percentage, substantially rewarding the cowboy crew with positive results for their labor.

Beaverhead Ranch artificially inseminated between 1,150 and 1,400 heifers each year for seventeen or eighteen years. We synchronized estrus using MGA—an orally active progestin—in the feed and time-inseminated them with no heat detection. Nathan Anderson, from the farm crew, contributed to our AI synchronization by diligent and precise feeding according to MGA requirements. We would AI all the heifers in a three-day window each year. Over the years, the average of heifers calving AI was 63% of total artificially inseminated heifers. In the highest percentage year, 72% were AI calves, and in the lowest year, 59%.

About this time, I brainstormed with Tom Elliot of the N-Bar Ranch, producer of some of the most functional Angus cattle in the world. Tom was known for thinking outside the box, so we had some stimulating conversations. I was still trying to fine-tune the most favorable bull genetics for use on our heifers. Calving ease was a must for bull genetics chosen for use on our Hereford x Angus heifers. While calving ease was critical, we also wanted to produce a robust, consistent, and desirable calf. Tom suggested I research a new Angus variation called Lowline Angus.

I learned that Lowline Angus cattle were developed in Australia and were now beginning to see use in the United

States. They were proven calving-ease bulls with a smaller frame, but greater capacity, than most modern Angus genetics provided. Due to crossbreeding heterosis, our cows were beginning to increase in physical size. Lowline genetics could provide a tool to maintain the moderate cow size we needed while getting more live calves on the ground.

We artificially inseminated a group of fifty heifers to a Lowline bull to test the new breed. The resulting calves were small at birth but possessed a lot of vigor. They remained smaller than our other calves and reminded us of the Angus cattle of the 1950s and 1960s, with a smaller frame, shorter legs, and thick-bodied with lots of capacity. Those traits fit nicely into a herd required to graze year-round with minimal outside inputs. There was one problem, however. Their short, squatty bodies did not appeal to buyers that wanted and were accustomed to buying larger frames.

We grazed some of those first offspring on small unused areas of the ranch, including areas that needed weed eating, until they were sixteen months of age, at which point we had them harvested. We provided the cut and wrapped beef to employees to test. We had the steers graded for quality at the livestock plant, and most graded Low Choice despite having had a diet of 100 percent grass and weeds. The beef was great, and in the following years, most employees preferred to receive Lowline beef when given a choice.

Our initial experiment excited me, and we then bred all of our heifers to Lowline bulls. The resulting calving ease was extraordinary. Our four-person heifer calving crew calved 1,150 first-calf heifers with an unheard-of 98 percent live-calf crop. Most cattle producers—especially in the northern tier—will never achieve such phenomenal results on a large herd of first-calf heifers. We recognized the crew for their once-in-a-lifetime accomplishment by increasing their pay. As an incentive reward, the ranch gave each of the four employees money to purchase a custom-made saddle.

Cattle buyers still scoffed at our squatty-body steers, so we put them on feed in Kansas and retained ownership. We used these same steers in a research project with Kansas State University, testing retention and recovery of the electronic ID implants in ears. They accounted for all but two out of 240 implants. One implant was lost, and a lightning strike destroyed the other.

The steers fed better than anyone expected, but were still small-framed. I was concerned the feeder would try to make them bigger than they should and feed them too long. The average live weight at processing was 1,158 pounds, with a large percentage grading Choice. Many had 0.7-inch backfat, which would usually result in high yield grades. Surprisingly, there were only four USDA yield grade 4s out of over 200 steers. The unexpectedly large 13.86-inch rib eye average was the reason for the excellent yield grade. Retaining ownership as these half-Lowline steers moved through the system was a good learning experience and furthered my opinion that these cattle could be a good fit for grass-fat meat programs. We continued breeding heifers with Lowline Angus until I left in 2011.

As with our initial crossbreeding experiment, we underestimated what those half-Lowline females could do when put in the cow herd. As cows, the half-Lowlines bred like rabbits, maintained body condition, and weaned an average of 51% of their 1,000-pound body weight *on grass only—no hay*. The cows were feminine with beautiful udders and were the most efficient cows I encountered in my entire career. We had about 500 half-Lowline cows when I retired. A few years after retirement, Sue and I purchased four of these half-Lowline cows at auction from the herd I'd developed to add to our small herd. They continue to be the efficient cattle we designed. Our daughter Kristy and husband Gregg Hoy in Kansas use a Lowline bull in their grass-fat meat business.

The first decade of 2000 achieved significant growth in ranch expansion with several neighboring ranch purchases. John Erb divested his Blacktail holdings, and we added the Brown Place and Middle Fork of the Blacktail, providing summer grazing for 800 more cows. When Bolings moved, we had the opportunity to purchase the old Selway Ranch in the Blacktail at Price Creek. Also known as the Chaffin place, this addition provided several thousand more grazing acres and a historic headquarters about twenty miles south of Matador Cattle Company headquarters.

We purchased the Jack Thomas Ranch in the Centennial Valley, which had good grazing for a few hundred livestock and a nice set of buildings. Next was the Clover Creek place purchase from the Hess family, which adjoined the Thomas Ranch, the US Forest allotment attached to our Middlefork purchase, and much of our existing Centennial Valley country.

Another addition was the Kent place in Price Creek, which connected our Sage Creek Ranch, the Blacktail meadow area, the Selway Ranch, and Blacktail Ridge country. The pioneer stage and freight road from Corrine, Utah, to Virginia City, Montana Territory, in the 1860s and 1870s traversed through the length of this property.

The final land acquisition during my tenure was the purchase of the old Wilson Ranch from the Duffner family. This ranch was in the Long Creek area of the Centennial Valley and adjoined much of the Matador's original ranch property. All these added properties were either adjoining or inholding properties to the Matador. They were also previously owned by some truly great neighbors.

From a land ownership standpoint for Koch Industries, Inc., blocking up an immense area of southwest Montana as the properties became available made sense. The uniqueness of the ranch became even more impressive, with over 300,000 acres in one block of open space, roughly the size of Grand Teton National Park. With the addition of these properties, we did not increase our staff, but rather, we increased the responsibilities and authorities of the four existing herd managers and the farm foreman. Each had one full-time support help and part-time or seasonal family members to accomplish their goals. They rose to the challenge, and not just for a couple of years. Until I left in 2011, they were still meeting and exceeding goals. We had only one personnel change in the herd managers.

Todd and Rita Sawyer and their boys started in the early 1990s and, in 2022, Todd is still running the Price Creek Division. Vernon and Vila Krug also started in the early 1990s, and except for a short stint in Kansas, Vernon is still managing Sage Creek in 2022. Todd and Vernon and their families have been the solid rocks of our crews through the years. Their wives, Rita and Vila, have contributed as much as any employee. Besides helping wherever needed and being support help for their husbands' remote jobs, they voluntarily took on the ditch irrigation of thousands of acres of meadows, which were fundamental drivers of profit for the ranch.

Other key longtime employees were herd managers Ivan Burch and Dave Nick and support-help Sandee Nelson. I do not intend to diminish the efforts or inputs of others but recognize these as having the most long-term effect on ranch success. They all grew in experience and effectiveness and were fortunate to be part of the ranch management restructuring, which coincided with Koch's move to MBM—market-based management.

In 2005 another support-help cowboy came to the Matador from Kansas. His name was Gregg Hoy. We knew him as Skip, and he had come to experience cowboying in Montana, which was quite different from his home country, the Flint Hills of Kansas. He and his cousin, Josh, had day-worked for Sterling Varner at times in the Flint Hills and developed an interest in the Beaverhead Ranch from Sterling's stories, I assume. It was quickly apparent that Gregg came to us with a different attitude than the average cowboy. He had genuine questions and sincere interest in how we operated, which was more like an owner than a hireling. I enjoyed spending time mentoring him. Skip was a good hand, and we were sorry to see him leave and return to Kansas in 2006.

2005 - Gregg "Skip" Hoy heeling calves on Copy - Montana's Matador

Twelve years later, Gregg and our daughter Kristy, who was in Moldova, reconnected. I tell him that I allowed him to marry Kristy as thanks for bringing her back to American soil. Seriously, I always liked Skip and could see their friendship developing on the ranch in 2005. He had the kind of integrity and attitude that was hard to find, and I could see his potential. Sue and I were happy when they married and started building their own future with cattle and horses on their own place in Kansas. Skip's day job is a tenured conductor for the railroad, and Kristy works independently as a dental hygienist. They're a great team, and I couldn't imagine a better husband for our Kristy. One of our favorite things to do in retirement is to drive to Kansas in the middle of July and help them brand and ship cattle. I would never have believed that this Montana cowboy would say such a thing, but it's true.

Similar to the early 1970s, when Koch introduced benefit packages to the ranching industry that included health insurance and company-matched 401 savings plans, around 2000 they introduced *"Roles, Responsibilities, and Expectations"* for each full-time employee. Attached to the RR&Es was the new "performance pay" tool to provide monetary incentives when employees exceeded their expectations. From a manager's perspective, these tools were potent. With the employee's help, we could establish RR&Es that addressed measurable production that increased profitability and gave me a way to ascertain the values of those contributions.

These RR&Es were not a list of tasks as most traditional ag operations would write, but instead, focused on what we wanted to accomplish within their responsibilities, such as reaching or exceeding a base expectation in their grazing management. From that point, it was up to them how they would meet that expectation. Incentive bonuses are a win-win for the company as well as the employee. Performance pay rewards and encourages employees to develop initiative and innovation that contribute value to the company.

While this system was one of our most significant accomplishments, we also realized that over time, some employees would reach a point in their new position where they may have matured or begun to stagnate with no obvious new opportunities in view. It happens in every human organization regardless of structure. I call it "The Cain and Abel Syndrome," a biblical reference to one brother who killed the other brother out of envy. Whatever was going on, I learned there had been accusations from unknown parties who appeared to be circling the wagons, then shooting in. HR, the Human Resources department in Wichita, had gotten involved and was increasingly controlling business, keeping us from turning this problem around. It would eventually contribute to my "retirement."

> *"Cutting someone off at the knees does not make the small man bigger." Ray Marxer*

Programs we started in the 1990s were continuing and improving as we practiced them. One of the major focus areas was safety. Charles told me personally that "safety, or lack thereof," would be the biggest threat to Koch selling the ranches. Safety attitude was the same for all Koch businesses. They genuinely wanted all employees of Koch to go home to their families each night without injury.

Safety was a top-down driven program and needed to be. Focus on safety was a paradigm challenge. Did we believe we could operate this ranch without accidents or injury? Were we convinced that all accidents are preventable? The way we perceived safety was the most difficult challenge for all of us as it would require a genuine change in our personal beliefs. Most agricultural producers would question if getting things done accident-free while working with uncontrollable elements such as weather and animals was even possible. Traditionally, agriculture excels in production and getting things done, but it does not have a good safety record.

But we did it. Our answer was "yes," thanks to an extraordinary group of people making an outstanding commitment. We had many programs that would help us along the way, several unheard-of in the ranching industry. Some of our tools came from programs developed by Dupont, a company that Koch viewed as the safest company in America. Many in business and society would avoid that challenge by doing or producing little to nothing, using the excuse that they could not perform the work safely. One inherent danger in the quest for an accident-free workplace, is that pressure on crews could cause a failure for individuals to report minor incidents.

The "excuse to avoid work" was not an option for us as we intended to operate a massive, sustainable ranch business for both profit and the environment. All three ranches accomplished a few years without a lost-time accident, and the Beaverhead went three years without a recordable accident—an incident requiring medical attention or prescription. Three injury-free years was a monumental achievement for any ranch. Imagine my disappointment when I was the one to break that streak when I injured an ACL with an after-hours, at-home incident with a new ranch horse.

Many of our safety efforts focused on horse safety and improved horsemanship. As I have mentioned before, the selection and development of horses and people were some of the most critical and challenging aspects of ranching. We continued bi-annual horse safety and horsemanship clinics at the ranch with world renown clinician, Mike Bridges, from the mid-1990s until I left in 2011. We built an indoor arena to help employees work on horses and horsemanship during the winter when more time was available. We also designed that barn with pens down one side that the crew would use for calving heifers.

Much improvement in horse and rider development occurred, and to help in the horse selection process, we started raising some of the horses needed. Some of the ranch-raised horses worked quite well, and others did not, just like the purchased horses. Through the 1970s, most training was on the job, and the purchase budget was $750 to $1,000 per horse. The ranching rule of thumb for large operations at the time was that a fair price to expect to pay for a horse was two and one-half times the value of a cow. The average cost we incurred for horses we raised, accounting for all those that fell out, and hiring a trainer for the first thirty rides was about $3,000 per horse that made it to a cowboy's string. We tried numerous sources and evaluation procedures to acquire horses that fit our needs. After I left in 2011, the ranch tried increasing the budget to $10,000 per horse and still had some that fit and some that did not.

Except for safety and management structure change, many top-down-driven ideas attempted at the ranches were not very successful. The prominent venture in the early 2000s decade, was hunting enterprises at the ranches. Brainstorming sessions with some of the large ranches in America at the NCBA convention showed us that hunting could be another revenue stream for the ranches. The Deseret Ranches were getting 17 percent of their income from hunting.

The company would make a big push at the Texas ranch since ownership was 100 percent private land and it was close in proximity to possible clients. They went all out with a high fence, a guest lodge, a television show, and an AI facility for deer. Ultimately, the returns did not pay for the costs, and the Texas ranch discontinued its hunting program. The Beaverhead Ranch had the challenge of sections of public land mixed with sections of private land, and state laws disallowing high-fence game farms. Consequently, recreational hunting was never attempted outside of the company.

There were several "suit" changes during the 1990s and early 2000s, most of which were very helpful. We had Larry Angell back as our Matador Cattle Company president for a while, and along with others in Koch, we were acknowledged and rewarded for the many innovations and stewardship accomplishments of the 1990s.

Larry retired a couple of years earlier than I did, and Sue and I went to Wichita for his retirement party. While visiting, I asked what made him decide to retire. Larry answered with some critical advice I probably should have considered more carefully. "We've reached the top of our field," he said, "and generally, the only way to go from there is down."

To me, "top of our field," meant "breakthrough advancements," and I knew we could still improve at a more incremental level. Only three years remained on a ten-year commitment incentive plan the company had offered me, and I determined to proceed for at least three more years. Sue and I both were involved with organizations and community groups that would affect the ranch. I was active on:

- Governor-appointed Private Land/Public Wildlife Council for six years

- Governor's Task Force for Cutthroat Trout

- Board of Directors for the Montana Stockgrowers Association

- Beaverhead County Fair Board for sixteen years

- Numerous other advisory boards, for which I continue to serve

Sue served on the BLM Resource Advisory Council for six years and was chairman when the BLM was in the middle of renewing its ten-year Resource Management Plan. The ability to appoint balanced committees to deal with specific issues effectively kept the council's advice reasonable for all interested parties. She also worked on the Sage Grouse Council.

We continued to host the World Wide Country Tour's "Big Sky Country" bus tour for a ranch stop three to five times each summer as we did for around twenty years. We shared our part of the country, ranch and family life, and local history. We advocated for beef, federal lands grazing, and the company. We loved sharing our story and meeting many great country-minded people from New York to California and everywhere in between. Sue would usually be tour director for one or two entire WWCT tours every year. Besides the Big Sky tour, she directed a Calgary Stampede tour, Fall in New England, Lewis and Clark tour, and others.

We were happy to be invited to join *Provider Pals* as a rancher representative for a few years. Provider Pals was a sponsored volunteer outreach program targeting inner-city junior high students without exposure to natural resource industries. The innovative founder was Bruce Vincent, a logger and industry advocate from Libby, Montana. Volunteer producers from varied natural resource industries visited large metropolitan inner-city schools who were interested. They presented to the seventh and eighth-grade students how they worked with natural resources and what they did for a living. Several students would earn scholarships to attend a Provider Pals summer camp and learning center in Libby, Montana, the following summer by completing follow-up projects and receiving a referral from their teachers.

Provider Pals was a good fit for Sue and me as we had experience sharing ranching and the western way of life, including slide show presentations and roping demonstrations. We have always had a strong passion for sharing with young people and found it very rewarding. The kids were interested in our cowboy life and got a brief glimpse into a way of life different from theirs. Two memorable cities we presented in were Washington DC and Little Rock, Arkansas. Other industries accompanying us on these trips were loggers, miners, commercial fishermen, and grain farmers. It was so fun to teach numerous kids at each location how to rope a dummy calf on a hay bale—provided by the resourceful Provider Pals staff. An unexpected treat for me was visiting one-on-one with parents with preschool siblings waiting to pick up their students at the end of the school day.

Provider Pals visit to Washington DC   Susan Marxer photo

I appreciated the incredible opportunities we were privileged to participate in and enjoyed sharing ranch history and stories. For any awards, we shared credit with the people and families who made Montana's Matador Cattle Company successful, and with Koch for allowing us to treat the ranch like owners. We never ended a presentation without honoring Jesus Christ, the source and sustainer of America's abundant natural resources. We were simply willing tools and thankful stewards of an oft-maligned industry. An aging and dwindling two-percent of America's population is responsible to be good land stewards providing food and fiber to the other 98 percent—many of whom believe food comes from the grocery store. The day America becomes reliant on other nations for a safe and dependable food supply is the day we are no longer "the land of the free."

There were always up and downs on the ranch and even more so in Wichita cycles, but life goes on. I'd been around long enough to observe how society had changed from fundamental, cultural, and generational standpoints. Being part of the vast, historical Matador Cattle Company was not a privilege we took lightly. We loved where and how we lived, the country, the cattle work, and bringing up our kids in a rare and treasured environment. When the ranch finally ran electricity into the Staudaher Cow Camp, it was the end of an era similar to transitioning to horse trailers when I first started my ranch career.

CHAPTER 15

Who Can Stand Before Envy— and HR

"No crime is so great as daring to excel." Winston Churchill

By 2004, Koch recognized many of the monumental strides we'd made on the Beaverhead Ranch. They offered to build a new manager's house for us since the existing one was old and had typical old-house issues. The ranch had upgraded most of the employee housing except for the three old frame homes—the manager's house, the farm foreman's home, and the little house across the creek. We politely declined the offer since we were comfortable and saw no need to go to that expense. Shortly after, the company approached me, desiring to create an incentive for me. They wanted a guarantee I would commit to continuing as manager for ten more years.

"What would you like to do when you retire?" the Matador Cattle Company president wanted to know.

"What we want to do is get a little place of our own, somewhere in this same part of the country, big enough to run a few cattle," I responded. "The only place the kids have ever known as home is this ranch, and whatever house we happened to be living in. It's important to us that we provide a real home for them to come home to and a place to bring grandkids. We'd like to stay engaged in the beef industry—along with our family."

We were a bit surprised when the company went to work right away, looking at a couple of places with us that might work. Neither one panned out, so they returned to the drawing board and came up with another idea. They offered us the opportunity to purchase an old, abandoned homestead in the new Brown Ranch property the ranch had acquired from John Erb. They would begin constructing a new house for us with a budget commensurate with the two properties we'd viewed. The company arranged an incentive compensation package with 10 percent of the value paid annually as a taxable bonus. If we invested wisely, we would be able to purchase the retirement property for the current 2004 price at the end of the agreement in 2014. They even gave us an option to retire in seven years, 2011, and still purchase the property if we came up with funds for the remaining three years. The company retained the right of first refusal if our family ever sold it. The generosity of that offer astounded us.

The company lawyers drew up the agreement and suggested we have a lawyer look it over, which we did. Our attorney advised that we refuse the reasons for termination of the contract. They were too broad in that we only had to be *accused* of wrongdoing, whether true or not, to void the agreement. The company lawyers refused to revise the contract, so we signed. We trusted that the company would continue to honor its intention.

We had a budget and designed a house with an attached garage. Adding the garage meant it would require sweat equity from Sue and me to meet the budget. The home was beautiful but not extravagant. We wanted the company to be able to use it as a modest lodge or home in the event we didn't want to live that far out after retirement. One of my closest friends, Dean Simonsen of DBS Builders in Dillon, was the contractor and did an excellent job.

Sue spent months on the computer designing the house and purchasing fixtures online with her own money to help us stay within budget. Sue crafted the kitchen backsplash with some of her favorite black-and-white photos on tile, did almost all the priming and painting, and finished all the interior wood trim. She had occasional help from our daughters, and a couple of friends from church who she paid herself. She cleaned and finished the colored stamped concrete we'd used for the basement, finished the doors, and purchased and planted shrubs and flowers. Sue's brother Ken, a flooring contractor, installed the laminate on the main floor to repay a favor to her.

The ranch paid for numerous small windbreak trees and bushes from the state nursery that Sue and I and our daughters planted and tended. Clayton and Nathan Anderson helped me take the makeshift top off the old homestead barn, reroof it, and make the old barn usable. We made a sealed tack room and rebuilt a couple of horse stalls. Those two also helped me build an arena that doubled as ranch branding, weaning, and shipping pens.

I helped with road and site development. In the hills behind the place, I found and gathered the perfect flat stones I used to build two large fireplace faces and hearths. We used a wood insert in the basement fireplace and a propane insert on the main floor. My great-grandfather was a master stone mason, and I enjoyed dabbling in his profession. The stones and mantels collected from the property and ranch made the fireplaces unique and special.

I crafted the main floor mantel from a giant juniper I'd found that had been struck by lightning and had grown horizontal. The three sawn sides revealed deep, beautiful multicolored wood grain. I made the basement mantel using a massive old jack-leg post with hand-drilled pole holes. The significance of that jack-leg was that it was a remnant from the early fence constructed by Poindexter and Orr more than a hundred years earlier.

All this work was secondary to my manager's job and branding. Sue was helping with branding and packing up our headquarters home.

The year we built the house and moved there, I contracted Colorado Tick Fever. It was a challenging year for me being sick and for Sue to put up with me feeling so poorly for several months. Sue was worn out. The tension with the unknown HR system workers continued, and Sue said she didn't know who to trust. She was "just waiting for the other shoe to drop—usually, when something sounds too good to be true, it likely is." She loved our house but guarded against getting too attached. She never talked about it or asked for help from anyone on the ranch. By then, the company president, who had been a great advocate for us and an encouragement, had been put to work elsewhere in Koch, so a new suit replaced him. The transition is typically a little rough when everyone is relearning everything again. The suit change made it more problematic due to our unusual agreement, which was getting lost in the shuffle.

We moved into our new home in the late fall of 2004. Since the girls were working in town, the ranch let them rent one of the old empty trailers at headquarters rather than drive from the new house. Our road wasn't suitable for car travel. We enjoyed having the kids come home from college for Thanksgiving. They brought a few friends, which made Thanksgiving extra fun and memorable. The girls and Clayton's college buddies spent one entire night around the kitchen table playing pinochle and eating pumpkin pie. Clayton and Nathan tried to get some sleep to get an early start for hunting.

Around 2009 we had yet another suit change, meaning I would need to start over to re-earn decision rights. This boss happened to be one of the muscle-flexing chest-thumpers. The suit drove up to the house with me on one of his first visits, interested and anxious to see what we'd done. On the drive up, we crossed the lush Blacktail meadows along Blacktail Creek and talked about the abundant population of moose and other wildlife that were regular visitors to the ranch's creek-bottom property, below the property where we'd built our new house. I could see the wheels turning. The Beaverhead Ranch had become popular with several company folks who enjoyed hunting, and he supported the idea of a recreational hunting enterprise.

On the next trip from Wichita, he walked into my office with my retirement agreement in his hand, saying we needed to talk. And talk he did.

"I have negated this agreement," he proclaimed, "and we're replacing it. I don't think the company ever meant for you to own land within the ranch. I don't think it's right that you or your family should be able to own that land and house."

He continued explaining the revision, trying to convince me that we were better off and weren't out anything. "We're still taking care of you. We've revised the agreement so you can live there as long as you want, and we'll take care of taxes and maintenance."

According to their new agreement, Sue and I could lease the property until we passed, but with restrictions. None of our family could live with us, continue the lease, or build a second residence. Our use of the land could not exceed his limits. He reminded me of the saying, "I'm from the government, and I'm here to help."

Another sucker punch—a second spit-in-the-face. In spite of the discouragement, we kept moving forward.

Next, the new suit changed our enormously helpful enterprise accounting system to a cash-flow system generally only beneficial to upper management for decision-making. For the first time in my thirty-seven-year career with Koch, my employment had become a job. The stress level was affecting my health, and I was considering early retirement as numerous red flags were starting to wave.

A significant red flag was the blatant hiding of a proposed switch in cow genetics from Kyle and me. The secretiveness was out of character for Koch operations. The Wichita HR flag got bigger and bolder when HR increased its involvement. Now HR and the current suit were letting me know the source of some of these communications. HR moved all hiring to the Wichita office. HR took away my ability to conduct personal interviews and warned about retaliation against the "source"—two disgruntled employees who would now be conducting all personal interviews for new hires. One observation from a seasoned employee was that it appeared "we were letting the inmates run the asylum."

One morning in 2010, while driving to work, I enjoyed a particularly great fellowship with God. I stopped my truck with Ashbaugh Peak ahead of me in my view. Gratitude for what God had done for me overwhelmed me. I'd had unbelievable opportunities I never thought were possible. He'd given us astounding success in every area of the ranch thanks to a core of solid employees who worked as well as a team as they did individually. He'd given me a wonderful wife and family and allowed us many golden years of growing up on the ranch. We now had our first two grandchildren, Braxton and Alexa, and this fantastic new home and a place to retire. While sitting in that truck alone and crying with genuine thankfulness, I told God that He could take all of this away from me, and I would still be grateful for His grace and mercy.

2009 - Ray Marxer heeling calves for branding on Skipper. Ashbaugh Peak in background. *Susan Marxer photo*

I continued for more than a year, frequently contemplating retirement but still making progress and ranch profit that the company rewarded with a substantial year-end bonus. In the winter of 2010, while in Wichita for a meeting, a group of us were at a dinner outing. The suit told everyone at the table that he thought I should run for president. I'm guessing he was impressed with how I prefer to seek solutions and resolutions to problems rather than escalate or cause conflict and controversy regardless of the entity.

A few months later, HR received another "Cain and Abel Syndrome" communication. The person(s) accused me of "concerns" that instigated a full but fruitless investigation of me based on false assumptions. I could have easily listened to their questions or concerns if they weren't so "afraid" to talk to me. As far as I know, the "whistleblowers" were always given the benefit of the doubt that they were acting in good faith, and never faced any responsibility for abusing and weaponizing the HR system. They had earned a free pass since any rebuke or anything perceived as negative would be considered retaliation.

One morning about a week later, the suit showed up at my ranch office, accompanied by an HR person—and a bodyguard. They came in with U-Haul boxes. The suit abruptly told me he'd decided that I could no longer lead the team of employees here, and I was no longer employed by Koch. I could not get a straight answer from him, nor did HR ever provide me with the written letter of dismissal I requested.

I was understandably upset that a wonderful career would end this way. That cold, corporate crew in my office could have handled their disgraceful dismissal of me in other ways more honorable for Koch and me both. God's grace was all over me. I remained calm—even friendly—as my brain tried to process what had just happened. I did not want to be where I was not wanted. So, the cowboy who'd been a loyal employee for thirty-seven years—more than half the years that Koch had owned the ranch, was told, "We will help you box up your personal things in the office. Leave your phone and keys, and we want your laptop. We'll drive you home."

The original agreement allowed us to retire as early as seven years into the contract, which we were. When I asked if I would be allowed to retire and put into effect the retirement agreement the suit had changed, the answer was, "Nope. That's done."

So the security guy drove me "home" in my ranch truck. "How long have you worked for Koch?" I asked him as I directed him up the road to get to our house. He seemed like a nice guy who was feeling a bit awkward.

"Just a few months." We were driving by the crew who was branding in the new branding trap Kristy had helped me build for the ranch.

I honestly wanted to be an encouragement to him. "Koch is a great company to work for," I assured him. "In the thirty-seven years I've worked for them, I've never known them to do anything as radical as what you just witnessed. That is not typical, or representative of Charles Koch's leadership."

The following day I was offered a severance package to basically ensure that I would not file a wrongful discharge and discrimination suit against the company. It wasn't even close to what they'd just taken from us—but it would help us get relocated.

After much prayer, and remembering what I had told God that day on the way to work, I knew this was God's way of testing me. I agreed to the settlement but requested to "retire" to the public. My concern was not as much for my image but what the community's reaction would be when they heard the company fired me before I could retire. We spent over thirty years advocating for and sticking up for the image of Koch, and did not want that effort to be in vain. I was the middleman in a no-win situation and had been considering retirement for some time. We liked the Kochs and enjoyed friendships with many company people. Most of all, we loved calling Matador Cattle Company "home" the entire time our children were growing up. The visiting corporate trio approved a "retirement" press release for the local weekly newspaper in June, 2011.

In 2021, ten years after I left the Beaverhead Ranch in my rearview mirror, Koch sold all three of their ranches. Montana's Matador Cattle Company sold to media mogul, Rupert Murdoch.

In the words of an old friend and neighboring ranch cow boss, Bob Birrer, who worked up the Ruby for years on the Snowcrest Ranch, "When they're done with you, they are done with you." That day, I committed to myself and my family that I would not spend the rest of my life looking in the rearview mirror. I still had too much to accomplish. My new obstacle needed an opportunity.

"If thou faint in the day of adversity, thy strength is small."
Proverbs 24:10

Once again, God had kicked a leg out of my three-legged stool. Removing the "job" leg meant I would be leaning on Him a lot more. The other two legs, faith and family, remained solid. God reminded me that He had me in that job in the first place, so we would have a good church to grow our faith and raise our family. Our kids had spread their wings, and it was time for Sue and me to establish our own borders—a place the kids could come home to, our grandkids could call "Grandpa and Grandma's," and a solid investment we could leave for a family inheritance. "Purpose" became the new leg to rebalance my three-legged stool.

God gave us a newer and larger home than the one we built on the Beaverhead Ranch. The house still needed some finish-work and a yard. There was a barn, and I put up pens, built an arena, and fenced the place. About half the property was in irrigated alfalfa and grass hay. Those were perfect projects to keep our minds off the recent past and prepare for our future.

Our new neighbors are fantastic. For the first time in thirty years, we lived next-door to actual neighbors who weren't employees. We're still growing in the same church where God planted us thirty-six years ago. We continue involvement in ranching with our own ranch services business for AI, consulting, and herd work, and running a few cows of our own. I've enjoyed conducting sustainable business seminars, mentoring young people in the ranching industry, and helping with beginner roping clinics and Rodeo Bible Camps for kids.

"And the rain descended, and the floods came, and the winds blew, and beat upon that house; and it fell not: for it was founded upon a rock." Matthew 7:25

A recurring theme in this book has been the principle of the three-legged stool for balancing most things in life. I have realized for a long time that essential principles in life originate from the Bible, but a primary truth did not occur to me until writing this book. I was taking a daily walk and talking with God when He revealed a concept to me that should have been obvious. The origin of the three-legged stool principle came straight out of the Bible, where it's known as the "Trinity" (three-in-one): God the Father, God the Son, and God the Holy Spirit. A simple illustration is: water, ice, and steam.

That three-part balance we apply to so many things in business and life is ingrained since we are created in God's image. Man is born with body, soul, and spirit. My personal life balance is now: faith, family, and purpose.

"And God said, Let us make man in our image, after our likeness:" Genesis 1:26

CHAPTER 16

Essential Partners—Our Cattle

"He causeth the grass to grow for the cattle, and herb for the service of man: that he may bring forth food out of the earth;" Psalms 104:14

This book would not be complete without a segment on the cows that provided the product for this ranch called the Matador. As stated earlier, the cow herd was straight horned Hereford breeding from 1951 through 1989. Most of the genetics for the decade of the 1970s were sourced in Canada. Marion bought many bulls in Alberta based on visual appraisal and pedigrees. Most of those breeders had birth weight information but little else as far as production and performance data. That was the challenge he faced in purchasing bulls to use on first-calf heifers as well. The outcomes of heifer bull selection would vary greatly.

As general manager, the responsibility for bull selection was one of my chief tasks. Fortunately, the beef industry now had more tools to assist in that selection. I also had the advantage of researching and determining what traits we needed to increase profit. My project spreadsheets revealed that even a slight increase in reproductive performance percentage would outweigh selection traits such as weaning weight or even carcass traits.

The three-legged stool concept defined the balance of the characteristics needed for our cow herd—reproduction, feed efficiency, and carcass traits. The best new tool was *Expected Progeny Difference* (EPD) records, a data-driven projection that measured known progeny performance of specific sires. The American Angus Association was the lead in developing EPD technology.

As with most new technology, the data was only as good as what users entered into the database. Solid data relied on the integrity of the breeders, especially early on, to produce accurate and repeatable projections. It was a great tool compared to what we had in earlier years but was also an opportunity for marketing manipulation early on. I recognized that accuracy could be variable until breeders had input several generations and years of information. With that in mind, I concentrated on traits like birth weight, weaning weight, and yearling weight to assure an increase in live-calf percentage, with adequate weaning weights, moderate milk, and yearling weights that were moderate, as they were an indicator of mature size.

I wanted to maintain a moderate-sized cow that would fit our environment and feed regime. We were breeding straight Hereford cows to Angus bulls the first few years, so heterosis would help us achieve some of these goals but could quickly increase cow size if yearling weight EPDs got too high. Reproductive performance could suffer if the milk EPD were to be too high given our feed source. In the Angus breed at the time, an eighty-one-pound yearling weight EPD would result in a 1,200-pound cow.

To ensure more accuracy in our selections, I chose bulls that fit my multiple trait parameters according to their EPDs, then looked at the actual performance of sire and dam two generations back. Going back two generations would allow me to see the bell curve variation and select individuals whose variation was slight. For example, an EPD record could show a bull with a birth EPD of "0", meaning tiny to no expected deviation. But, that bull may have had a sire with a +4, a significant variable for increased birth weight, and a cow with –4, a significant variable for lighter birthweight. The variation, in this case, would be considerable and one to avoid.

Later years would see more traits included in EPDs, such as rib eye area and marbling, that would help in improving the product for our customer. A balance of EPD traits is still an essential factor in designing or improving both profit and product. Concentrating primarily on profit may reduce the value of the product to the customer.

The feedlot and packer segments mainly focus on profit. Generally, cow-calf producers have relied too heavily on production, equating production to profit. By intensely focusing on certain carcass traits and feedlot performance to meet the wants of the feeder and packer, the cow-calf sector has caused the average physical size of beef cows to get too large to be profitable.

Production and profit are not the same.

Balance is key to producing value for the customer and profit for the producer. I can't repeat these three factors enough in regard to the three-legged stool for success in the cattle business:

- reproduction
- feed efficiency

- carcass traits

In 1991, these were the tools I used to select replacement bulls from Angus sales around Montana. Through trial and experience, I'd learned a great deal about which Angus lines would fit our big-country southwest Montana environment and which ones did not. I managed to find a unique calving-ease Angus bull at the Blevins Angus Ranch in Charlo, Montana. I found *Ambush Bando 01* by entering my EPD parameters using the Angus Association's database. "O-One," as we called him, came out number one in the entire country for yearling bulls.

I quickly contacted the Blevins Ranch and was able to purchase the bull at a very reasonable price. I would learn later that I beat Roy Wallace of Select Sires by a couple of days in purchasing him. Shortly after that, Roy tried to buy the bull from the Matador to put in the Select Sires AI stud catalog, but I declined. We used Ambush Bando 01 for several years to AI (artificially inseminate) our replacement heifers and then leased him out as an AI stud. O-One made a colossal improvement in our cow-herd for fertility and calving ease.

When O-One's daughters came into breeding age, we started using calving-ease Hereford bulls on them. We tried artificially inseminating them to Polled Hereford calving-ease bulls but found after three years that they had too much frame and insufficient capacity to fit our environment. We tested those cows against their ranch contemporaries bred to Hereford bulls, and they did not breed back good enough in the big, rugged country where they traveled and grazed—our environment.

We discontinued the use of Polled Hereford bulls as soon as possible. Mr. Koch always recommended, even provided, good management books with solid concepts applicable to any business. I'd read one about manufacturing titled *The Goal* that made the point that the earlier you could identify and eliminate noncompliant components in your product, the more profitable you would be.

The crossbreeding design we developed specifically for our cow herd and environment had several essential requirements. Moderate size, which we determined to be 1,100 to 1,200 pounds, was imperative. The cow needed to be a good grazer, have a good disposition, and bring in a good calf every year.

The F1 Black Baldy cow—the first cross between a Hereford cow and an Angus bull, would best fit those needs. Due to the high culling rate in our straight Hereford cows for eye and udder problems, I needed to design a system that would produce as close to an F1 cross without maintaining a straight Hereford cow herd. The Hereford provided good profitable traits that we did not want to lose, so we would use bulls to maintain that Hereford influence. I had witnessed many crossbreeding decisions that may have benefited from heterosis, but there wasn't a planned breeding strategy or design. Those producers ended up with a mongrelized herd of cows whose only common trait was that they had gotten 300 pounds larger and less efficient. These were our results:

- We would breed an F1 Black Baldy female to an Angus bull resulting in a three-quarter Angus x one-quarter Hereford.

- Those three-quarter Angus females would be bred to Hereford bulls all their life resulting in a three-eighths Angus and five-eighths Hereford cross.

- The three-eighths Angus x five-eighths Hereford females were bred to an Angus bull all their life resulting in as close to an F1 Hereford x Angus as possible without having a straight Hereford cow.

This system worked well for over twenty years providing acceptable profit and production while maintaining our desirable cow herd. Later, we would add another component, using Charolais bulls on what we called our "terminal herd." This herd was an exception to our standard practices and came about due to a severe late-spring storm.

The three-day storm that struck in the middle of our concentrated calving season resulted in a 3 percent loss of calves due to the storm's severity. Eighty percent of the calves in this large cow herd were born in the first twenty-one days— about 230 were born per day. The cows lost their calves due to weather, not reproductive problems. The ranch had already invested wintering costs in the cows, so we ran them over as dry cows (no calf) and bred them to Charolais bulls with the intent that the ranch would feed out all resulting calves.

The Charolais cross worked well and produced a very desirable meat product. We continued the terminal herd and added to it by keeping some drys over every year, reducing the need for expensive replacement heifer development. The herd built up to 1,000 cows before we reduced it by shipping around 400 of those cows to the Kansas ranch.

Interestingly, those cows never performed as well in Kansas as they did in Montana. Numerous cattlemen had told me that cows moved from south to north can acclimate but moving from north to south affects production, which appeared to be true in this case. Much of the time, acclimation depends on the breed and elevation.

We introduced one other cross for our replacement heifers—Lowline Angus bulls. This Lowline Angus cross resulted in high percentage live-calf crops and half Lowline x half Black Baldy cows. These Lowline cross cows matured at 1,000 pounds, and they weaned 51% of their body weight compared to our main herd, which averaged 1,160 pounds and weaned 44% of their body weight. These half-Lowlines could fit in a grass-fat meat program very well and were the most efficient cows I would encounter in my career.

To have cattle with balanced traits and consistent herd uniformity, I purchased many closely related sire groups from all three breeds. We sourced these bulls from several different producers. When we first started using Angus bulls in mass, we arranged with Gardiner Angus of Ashland, Kansas, to raise bulls of specific sires and dam lines. We gave them our desired pedigree a year and a half in advance, and they produced them for us. This arrangement worked well, and Henry Gardiner and his sons helped us a great deal in establishing our core crossbred cow herd. Later we would use other sources for the Angus replacement bulls.

The Angus sires we used extensively were DHD Traveler 6807, Ambush Bando 01, EXT, Right Time, and Future Direction. Future Direction was a bull that fit our balanced needs better than any others up to that point. We turned out ninety-one sons or grandsons of Future Direction on our three-eighths Angus x five-eighths Hereford cows one year. Later, researchers determined that this sire occasionally produced stillborn calves due to *curly calf syndrome* caused by a simple recessive gene—meaning both the sire and the dam had to contribute. Because we were breeding the bulls to the Hereford cross cows, we never had any problems.

From 1991 until about 2004, I had an easy time purchasing bulls with the traits I was looking for at a reasonable price. I was looking for moderate-framed bulls with calving ease and decent growth traits while most everyone else chased growth. I don't recall exactly how many hundred bulls I bought in that period, but the ranch spent an average of $2,100 per head. We generally replaced about fifty or sixty bulls each year and turned out around 330 bulls into the herd annually. Around 2004 or 2005, many producers realized that their cows had gotten way too big and consequently started selecting for moderation. That discovery by other producers almost doubled the price for the bulls I'd been buying. The increased demand sent average prices to $3,600 to $4,000 per bull.

The Hereford replacement bulls that fit our balanced trait requirements were more challenging to find. For example, when we started breeding three-quarter Angus heifers to Hereford bulls, I needed seven bulls for cleanup after artificial insemination, which on average results in a 65% pregnancy rate. I thought I would try Polled Hereford bulls to eliminate dehorning the baby calves. I could only find three bulls in the entire yearling Polled Hereford registry that would fit my EPD criteria, and two of them were east of the Mississippi.

Todd Sawyer - Clover Creek Hole on trail to Centennial Valley

Consequently, I decided to use horned Hereford bulls and still found them difficult to find. I was fortunate to find two Hereford breeders who had not chased the growth fad and had concentrated on fertility and moderate size, our most important profit drivers. Those two breeders were the Duncan Hereford Ranch outside of Joplin, Montana, and Jack Turner of Sayer, Oklahoma. They worked with us to provide what we wanted in Hereford bulls, some polled and mostly horned for many years. Sire groups that worked well for us were DH Yampa Dominator, CSU Aggie 3203, NO93, and L617.

When we started our terminal herd that would use Charolais bulls, I researched many breeders and found that Chuck Stipe and his family from Charlo, Montana, had what we needed. We expected the terminal herd cows to calve on the range with very little assistance, making calving ease a primary concern. Not wanting to add the possibility of additional calving stress to my crew's workload, I calved those cows myself the first year to test the results of using Charolais bulls. Happily, the ease of calving those cows surpassed my expectations. The terminal herd crossed with Stipe Charolais bulls worked well for many years.

The integrity of the bull breeder can heavily influence the success or failure of genetic selections. The breeders mentioned understood our program and genuinely wanted to provide the best genetics for our specifications. They dealt with honesty and integrity in their genetic product, and they all played an integral part in our cow herd's productivity and profit.

Like our longtime crew, I consider the cow herd rock stars. We gave them the tools they needed, and they went out year after year and met or exceeded our expectations. Because these cows worked so well for us and we could raise a replacement cow cheaper than most, we tried to sell bred heifers for a couple of years.

The Black Baldy heifers were a consistent lot that weighed about 850 pounds as a bred heifer in the fall. They were not big enough for most buyers at that time, as they wanted a 1,000-pound heifer that would turn into a 1,200-pound mature cow. That ain't going to happen. We have seen the result of those 1,000-pound heifers turning into the over 1,300-pound cows most prevalent today. We quit trying to sell bred heifers and learned that we could not go out and buy the kind of genetics that fit our environment and management strategy for sustained profit.

Our product's consistency and predictable performance, along with large numbers, made marketing our calves easier. Customer feedback and retained ownership to the rail supplied information needed to tweak our genetics and health programs for more profitable results. The consistency of our calves allowed the Koch Beef group to use them in comparing custom feed yards for cattle placement. In the 1990s, we marketed our calf crop internally to Koch Beef until it dissolved.

Steve Christiansen was an order buyer and marketing innovator, who would market our calves for many years, and get us feeding performance and carcass data needed to monitor our product for customer acceptance. The major marketer for all classes of our cattle through the years would be John Erb. He bought many thousands of calves and cows. Until the last couple of years I was at the ranch, I negotiated the selling of many millions of dollars' worth of cattle to John with a handshake. John was the toughest of traders, but once we made the deal, it was a done deal. We never had a problem.

The consistency of this cow herd made it easy for me to make projections, whether it may be feed needed or weaning weights and head counts for shipping months in advance. It also allowed me to make a projected budget of expenses and revenue for this extensive operation in less than a day. I recall one year that my projected average weaning weight for over 6,700 calves was within two pounds of actual weight. Consistency also made organizing weaning head counts and weights for trucks relatively easy. In all the years of shipping that I was involved in, we never sent an overweight truck away from the chute and never had a stray that we had not found and separated before the brand inspector.

I am thankful that God created cows and other beasts of the field to provide food for mankind. But even more, I appreciate that for generations, people have had examples and opportunities to observe the miracle of birth, the cycle of life, and the passing of life for a purpose. Children who have witnessed the ongoing cycle have a deeper understanding and greater respect for life than so many who have never had the opportunity to live a basic life on the land.

CHAPTER 17

Horses—our Humble Heroes

"A man that don't love a horse, there is something the matter with him." Will Rogers

Ranch Horses as a title does not begin to say enough about the cowboy's most essential partner—the cow horse. Like people, horses each have their own personality, strengths, and quirks. Some are extraordinary, some are stubborn and hard-headed, and some are flat-out rebels that don't make the cut. They are all strong, silent, and enduring team members in their own right, regardless of their distinct and particular disciplines. Most ranch horses will work their hearts out when treated with respect, care, and proper training. I believe horses were essential to man's life and progress throughout the centuries and one of God's greatest gifts to mankind.

1987 - Staudaher Horse Corral, Ray Marxer

I have said many times that if there were such a thing as reincarnation—which there is not—I would not want to come back as a cavvy horse on a traditional big outfit of the past. With the rapid turnover of crews, a horse had to adapt to many different riders. What one would train them to do, another might dislike or punish them for doing. The ranch horse was a durable servant who endured many years and hard miles. I hope these words spark memories in the hearts of cowboys and cowgirls that rode some of these horses, and bring to mind others who deserve a place in a book as well.

A cowboy's contribution to mankind is greatly diminished when he's afoot. Without cow horses, there would be no cowboys. On a ranch the size of Montana's Matador, cow work would be impossible without good horses and good riders to partner with them. Horses, like most cowboys, spend their lives in virtual obscurity to people outside of the specific ranch environment. Some of these hard-working ranch horses deserve a book or, at the least, a shared story similar to *Smoky the Cowhorse*, the classic Will James book. I've mentioned a few horses in this book already, but I can't complete a memoir without remembering the horses.

> *"The horse. Here is nobility without conceit, friendship without envy, beauty without vanity. A willing servant, yet never a slave."* Ronald Duncan

Many of the ranch-raised horses on the ranch when I hired on in 1974 were out of Percheron studs used on Thoroughbred mares. Big, athletic, and tough, some of these 1,400-pound horses had what appealed to me as the perfect confirmation. The best ones were big and stout with good bones, straight and sturdy legs, and big enough feet on all four corners. I can't emphasize enough the importance of good, round, hard, and preferably black feet.

The ideal ranch horse needed a big muscular hip, good withers, and a strong back line that wasn't overly long. The horse would have a broad chest and a big-cinching girth that supported strong lungs and heart and allowed for a secure, tight cinch. The neck should set nicely into shoulders and not be a long pencil neck or too short and thick. An attractive head ideally has a good set at the poll and a kind eye. This horse would stand at about 15.1 to 15.2 hands.

Cow boss Tom Griggs had an eye for that ideal horse, and when I hired on in 1974, he had several in his string—"Roany," a red roan he used a lot for dragging calves, Bisco and Blue, some of the first ranch horses I had the privilege to ride, and two of his own, a couple of versatile, well-broke palominos, Nip and Tuck. Other good cavvy horses were Johnnyco, Nugget, Pronto, Matador, Jipp, and Zeb. Zeb was the last of the earlier ranch-raised colts to enter the cavvy in the mid-1970s.

Some other horses at the time were Max, Todd, Buttons, and Luke, a My Texas Dandy-bred buckskin I had at Sage Creek. He was the best all-around horse I've ever ridden and one I felt comfortable letting Sue ride until I had gauged her level of horsemanship. She'd grown up riding bareback on Welsh to quarter-pony-size horses in mountain country and was pretty fearless. Sue would be the first to admit that riding full-size saddle horses at a long trot in the big sage country, learning to rope, and working large herds of beef cattle took horseback riding to a new level. As a teenager, she'd gotten a taste of riding in big, open ranch country on a couple of visits to the Mellott Ranch in Moorcroft, Wyoming, which only stoked her dream of marrying a cowboy and living on such a ranch.

Early one winter at Sage Creek, I was doctoring a set of weaner calves on Luke. We'd slipped quietly up behind a calf that needed attention before I goosed Luke a little to make a quick and easy catch. Just as my loop was about to settle over the calf's neck, we hit an icy patch. The calf slipped and fell, and so did my horse.

Luke went down on his side so quickly my right ankle was mashed between the saddle and the frozen ground. I was clear of the stirrups but knew I'd broken my ankle. Luke was a little boogered when he got up. I had tied my rope on for calf roping, so I grabbed the rope, and Luke faced up. He was still stirred up and kept working the rope, keeping it tight—just like I always asked him to do. He was bothered by me floundering around on the ground, and if I didn't stand up, he was not going to stick around. "Whoa boy, whoa...easy kid...you're okay." I kept talking low and easy to him, wishing I could say I was okay too.

I got up on my good leg and stood there, still talking to him. I kept a hold of the rope and began trying to hop to him. Luke wasn't sure what I was up to, but eventually, I got him to stand still and allow me to climb back on and ride home about a mile and a half. My then-wife took care of Luke, then helped me into the truck so she could drive me the fifty miles to the doctor to get my leg cast.

A few years after Sue and I were married, Luke would fall with me for the last time. It was a cold spring night, riding through the calving cows. In one instant, without any warning whatsoever, Luke died under me from a massive heart attack before we even hit the ground. The loss of solid, faithful Luke was a hard pill to swallow. I was glad his death was instantaneous while doing what he loved, but man, I missed that horse. There've been a few horses over the years that I've wished I could clone. Luke was the first and maybe even the best.

Alpo—the little white horse cut to me my first day, retired, and spent his older years as a patient teacher for Clayton and later for Anna. I had Frank in my string for a long time. Frank, a classy-looking bay with a white strip down the center of his nose, was a handy cow horse and lighter built than some of my others. He made a tremendous step-up horse for Clayton when I wasn't using him.

When Tom first hired me, the using cavvy had over a hundred horses. Undoubtedly there were a few renegades who mainly packed salt. One we bought in later years was named Comanche. He could be a pretty decent horse if treated with respect. But Comanche had no patience for being retrained every time a new rider climbed on him. He would quickly teach the overbearing rider a little respect. Consequently, we didn't assign him to anyone's string but kept him around since he was a very effective teacher should the need arise.

Through the years, we used many sources for replacement horses, some with good results and some not. I recall a couple we got from Tom Davis at the Rock Creek Cattle Company in Deer Lodge. One was named Speckled Bird, and another Badger, the toughest horse I ever rode. While Tom Griggs was still the cow boss, we got a truckload of horses from central Wyoming. Most of these came "started," which amounted to being saddled and ridden out of the corral. When they made it back with their rider, they were ready for a cowboy's string. One of those was a palomino I called Sage that I had to start over. He remembered his first start, and it took a month of building trust before he would even stand for a halter. I got him to where he was usable, and he later returned to the cavvy at headquarters.

Ron Weekes and Marion Cross would source a few horses from a trader in Idaho. One memorable horse in this group was ID, short for Idaho Doc. He was a big gray roan and the biggest Doc Bar I would ever see. Ron rode him a long time, and then I took him into my string when Ron left. Sue drug a lot of calves on ID, then, since he was aging, Kristy rode him for several years after losing Old Yeller when she was eight. Jake was a big and stout gray roan that was gentle and got along with almost anyone. Jake and Happy Appy, the wrangle horse, were used occasionally by other families who wanted to ride.

Another memorable horse from my Sage Creek years was Bally, who appeared to have come off the desert of Idaho or Nevada. His confirmation and the size of his head suggested that he had some Shire in him. While he wasn't much to look at, his big head was full of brains. Bally was rangy looking, and he had a bluffing personality. He was strong and was my choice for working bulls. I could confidently rope a bull that was refusing to cooperate on Bally. He would happily drag a bull on its side if need be. He had a lot of "bottom" and was the best—but not my pick—for big, rough circles or the demanding job of driving steers up the mountain to the Ben Holt. He had a gait rougher than riding a bicycle down a railroad track.

In the late 1970s, we started buying most of our replacement horses from Bob and Lee Douglas of Sheridan, Wyoming. They had a horse-leasing business for dude ranches and had access to many horses that would work for us but possibly not for dudes. Bob and Lee were our source for more than fifteen years and supplied us with eight-to twelve-head annually. Bob told me he figured we bought a little over 120 head from them through the years.

There were too many to list here, and I will miss many memorable ones. Bill, Dick, Hawkeye, Hondo, Runs-like-the-Wind, Grande, J-bar, Charlie Brown, Sailor, Frank, Bomber, and Peter Paint, who would be a mount for Jim Bishop and, later, for our kids through their teens. Copenhagen, Red Wolf, Black Wolf, Blue Wolf, Macho, Sunny, Stub Ears, Teddy, Tommy, Mule, Plowowski, Gilligan, Sterling, Tex, Sox, Hollywood, Little Rock, Smoke, Amigo, Comanche, Powder, Ears, Bow Tie, Shoshone, Salty, Andy, Charlie—a pinto with a counterfeit streak, and many more. We had to find other sources when Bob and Lee quit their horse leasing business.

We bought some from private sources—Poco, Billy, and Pepper—and one memorable little "Doc's Handy Dan" horse we bought at the Bozeman Winter Fair. Danny was an intense, tough little cow horse that could cut with the best of them and switch ends almost faster than you could ride. He was tough as nails with a huge heart and a kind demeanor. Danny was in my string. Sue rode him some, and Clayton rode him a lot. Paying attention was an absolute necessity when riding Danny because he was so quick and responsive. We had to be careful on long demanding jobs at which the little horse with the big heart excelled. Danny would die before he would quit. Sadly, we lost him too early to colic.

In the late 1980s, the ranch began buying horses at BLS horse sales in Billings, Montana. We got some great ones there. Some were young grade horses that were diamonds in the rough. Two of the best were Buddy, who went to Todd's string, and Question, who was in Sandee's string.

Big, Copy, Casey, and Vegas came from BLS, as did Pockets—a tall four-year-old black-and-white paint from South Dakota. I rode him for several years. Pockets was a friendly horse and quite cowy. But I don't think he'd seen many creeks. It never failed. He would balk at every little foot-wide stream and fret and fuss, then if he could get away with it, he'd gather himself up and make a six-foot leap to be sure he cleared it. I thought I was being nice to Kristy when, in her late teens, I let her ride Pockets to help move cows on the bench. That was until I was watching as she went to step him across one of those little streams, and instead of stepping, Pockets made a giant leap. Kristy is pretty handy and rode the jump fine—but he didn't stop. The dirty bugger just kept on high jumping after he made his first landing and bucked Kristy off.

We bought several horses that didn't pan out. If we made a wrong choice, we sent them back through the loose horse sale and told potential loose-horse buyers all we knew about the rejects. We did not want someone else to get burned even though we might have lost money.

The selection and development of ranch horses was not unlike the selection and development of the cowboys that partnered with them. Horses and horsemanship were crucial factors in our safety program. Most of our ranch accidents throughout the years were related to horse incidents, and it was increasingly difficult to find help with enough experience to recognize dangerous situations. We had to start buying horses more mature than the four- or five-year-old horses we looked for in earlier years.

When Sue and I went to Billings horse sales, part of our strategy beyond picking out potential horses from the catalog was checking their confirmation unsaddled in their pen. We'd watch them ride around before the sale, then decide which ones we wanted to check out further.

Sellers relate to potential buyers by quick visual appraisal. I looked like a seasoned ranch cowboy. Sue looked like an average middle-aged lady looking for a pasture pet she could trail ride occasionally. The seller would immediately assume I knew what I was doing, but if Sue approached them and requested to ride their horse, they would either figure the horse was safe enough for her, or that she might do something to cause the horse to show its true colors—in which case they wouldn't let her ride it. We'd cross that one off our list.

If they were happy to let her climb on and ride, I would be at the fence watching as she mounted and put it through some easy flexing and responsiveness tests. Then she'd quietly ride around the arena, testing willingness, responsiveness, attention, and gait transitions. She'd test the whoa, go, and backup and see if the horse would give his ribs and move his butt around off her leg. We'd rate it with stars and only bid on horses that appeared safe enough and responsive enough for Sue—meaning they would be safe for most any cowboy who ended up with the horse in their string. In later years a couple of the guys would go with me. We learned early on that purchasing horses was much like buying bulls. It was often more important who you were buying from than what was being offered for sale. We probably bought more than a hundred horses through BLS.

We still needed to do a better job of training our employees. The ranch only had one critical accident in the late 1980s, a few years before safety became a key driver of ranch success, but it was one accident too many and was 100 percent avoidable. An inexperienced hired hand at Sage Creek caught his leg in an open halter hanging off his saddle when he dismounted. He fell backward, and the saddle on which he had failed to tighten the cinch, pulled over on the horse's side, which caused the horse to jump sideways. The guy was jerked toward the horse, which caused a spooked runaway with the guy spinning through the sagebrush behind the horse on the end of the lead rope.

It was one of the scariest wrecks I've ever witnessed. By the time we cornered and stopped the horse, the guy's leg was badly broken with a compound fracture piercing his skin. Otherwise, he was okay. Those mistakes could have killed him. That was a wake-up call about selection and development, and training and safety. At least we all had annual first-aid training and knew how to stabilize him while someone drove the five miles to a phone and called for an ambulance. The ambulance met a helicopter to airlift him to a hospital in Idaho Falls.

When we decided to try raising some ranch horses, we bought five or six mares through BLS. They mainly were old foundation bloodlines that would produce good-sized horses with good bones and feet. Sue and I had purchased a small black double-bred Cozmo Jet stud for ourselves and would breed those mares to him for several years. Clayton, who nicknamed the stud "AJ", started and trained him. Together they would win Reserve Champion Novice Snaffle Horse as well as Top Horse and Rider in the stock horse show at the Beaverhead County Fair as a three-year-old. We raised some good horses out of those matings and some that did not work. We culled some of the mares and learned that early handling of the foals significantly impacted their later usefulness and compatibility.

Selection and development of ranch horses was probably more challenging than employees, and that hasn't changed. In retrospect, we might have all had more success developing horses if we had written RR&Es for each horse in our string. Some of the good ones we raised were Sport, Flash, Poquita, Wiley, Sarge, Mr. T, Skipper, Ginger, Rachel, Dandy, and KC. Some of these were full siblings and exceptional horses.

Rachel, one of the early fillies that Clayton started, was a versatile black mare that made it into Jake Butte's string. Jake was Todd Sawyer's support help on the Blacktail side, and his twin girls rode with him often. The girls rode one of Jake's personal horses—and Rachel. After Buttes left and Rachel started getting some age, Jared Pulliam from the farm crew was able to provide Rachel a retirement teaching his two boys to ride.

Sport was trained and ridden by Ivan Burch, who showed him successfully in many stock horse shows. Flash was a full sister that Sue trained and rode. Sue was able to take Flash with her when we left. Flash, similar to Sport, was very cowy and very responsive. She and Sue were a great team that developed a bond of total trust that comes from working with a colt into their mature years.

Skipper would be the colt I would train and ride through maturity. He was one of the earliest colts that Clayton started under saddle. Like Sue, I was able to take my horse, Skipper, when we left the ranch. What a legendary horse he's been in our local circles. He's one of those once-in-a-lifetime horses I wish I could clone. Built like the old-style horses I described at the beginning of the chapter, he weighs over 1,400 pounds, is a solid, cowy, dependable ranch horse that's gentle enough for anyone—including all the grandkids. Skipper's athletic enough for me to win a couple of saddles team roping on him, and has carried grandson Braxton to several trophy buckles in youth rodeos.

One of the unique things about the colts out of our little black stud was their ability to trot. Coming from a cowboy who has covered over 130,000 miles horseback while doing his job, the trot is the most efficient gait of all—especially the long, ground-eating extended trot he passed on to his get.

We had several personal family horses through the years that served us well. Our family used most of them on the ranch. Peso was the first horse I ever started and trained. Sis, CB, and Dave—the bronc that topped the Miles City Bucking Horse Sale—those three were part Thoroughbred. Shadow, a Rocket Wrangler-bred gelding was a fine-looking dark bay team-roping head horse for many years and the one I used in PRCA rodeos. He hated sage brush around his legs, and since he was my competition horse, I didn't use him much in the hills.

Nugget was a big Hancock and Three Chicks-bred buckskin which was quite versatile. He didn't like anyone pampering, petting, or loving on him, but some of that was bluff. Nugget was actually quite gentle as he matured. He always had his eyes on his surroundings, though, ready to shy at a horse-eating rock, a startled bird, or anything else that flopped or flew.

Koko Commander was the first colt that Sue started and trained. She got him as a yearling from Ted Harrison, took him to a clinic there, and went on to show in a few stock horse shows even though she hated showing. She liked the results from the consistent work it took to get a colt ready to show. Kristy used Koko for every event in high school rodeo, including cutting. Barney Commander, a full brother to Koko, replaced Shadow as my heading horse. Barney enjoyed ranch work, and the whole family was able to ride him. Both he and Koko had powerful engines and could drag calves like there was nothing behind them.

Drifter, Sue's replacement for Koko, carried himself high. Sue always said he was like riding a pogo stick out through the sagebrush. He had a lot of "cow", and a person really needed to be riding if trying to turn something back in the pasture. Sue accidentally grabbed him with her spur once on a hard turn, but since he jumped straight, all she lost was her hat. Drifter was a favorite for the Wire Field brandings because he was one of the strongest-pulling horses we'd ever ridden in a branding pen.

Ray Marxer on Drifter and Kyle Hardin heeling calves for branding.

A couple of great Harrington horses were Al—and AJ—the solid black stud we bought from Harringtons as a yearling. He only stood thirteen-one hands, but he never knew he was short. I loved riding him, and could out trot anyone I rode with. Kristy used AJ a lot for dragging calves and for helping me in the feedlot. Sue used him quite a bit during the course of our ranch services business after we left the ranch. AJ was the sire to our favorite using horses, two of which are still in my string—Skipper and Poncho. The last colt he sired for us, Pistol, is in Clayton's little herd. The mare was Poquita, one of the first ranch foals out of the new bred mares. Clayton started and trained Poquita, and was able to buy her when he had to leave the ranch.

Al'e Bar Isle, the second Harrington horse, was Clayton's 4-H yearling to five-year-old horse project. Al was a fun, personality-packed colt that was a great calf horse, but was gentle enough to teach Clayton's kids to ride years later. He was Doc O Dynamite-bred, and black with a crooked white stripe down his face. Alexa, Clayton's daughter who rode him during his final years, called him "Albert."

Kristy had a 4-H yearling to five-year horse project, that she bought from Farrell Wheeler. Taz was a tall, kind, sorrel horse who she made into a dependable, versatile, all-around ranch horse. I traded her out of Taz with a 1986 Ford Taurus when she left for college. Unfortunately, we lost Taz to colic during spring branding. The only bright spot was that Kristy's Taurus lasted until she had a good job to replace it.

Finally, there was that special, gentle Old Yeller, the buckskin horse that Tammi Huntsman retired to toddler, Kristy, who rode him until she was eight. He genuinely loved little girls and would never go faster than Kristy could ride. She had to wear gloves though, because Yeller had a bad habit of dropping his head to eat, pulling those leather reins through her soft little hands. Yeller always took care of Kristy by watching where he stepped. Kristy couldn't force him into a bog or wire if she tried.

One of the most touching examples of compassion and caring for others I have ever witnessed in my life was not between people. We ran all the horses we used as a family in one pen since they got along well, and some were getting old. Animals are not always kind to the aged.

Early one morning, while doing chores before school, Clayton discovered that Alpo, the very aged white horse Anna still rode, was injured and bleeding. He got Sue and me for help, and the girls came too. What we saw when we got to the corral was distressing. Alpo, the old "king of the corral," was sadly standing off in a corner and bleeding from a compound fracture from a high kick to his front leg.

What thoroughly amazed us was the sight of his old compadre, Peter Paint. Peter was standing guard over Alpo, running the other horses away. Then he would return to Alpo and lean against his shoulder to help hold him up. Peter Paint continued his vigilant care until we could get the vet to the ranch to put Alpo to rest.

What a heart-warming expression of kindness and compassion Peter Paint demonstrated for our kids and us as parents. It was never easy losing a faithful old friend and partner, and Peter Paint was a stellar example of standing up to his crowd of peers who wanted to injure that good old white soldier further.

Peter Paint showed the same compassion for Koko when he was dying from an internal condition he'd developed while healing from severe wire cuts. Months earlier, motorcycles roaring through the ranch yard on the county road, had spooked the colts and Koko, their baby-sitter, in the roadside pasture behind the barn. Clayton discovered Koko's injuries when we got home from church, and someone from the trailer court across the road told Clayton about the motorcycles and the panicked horses.

Nursing Koko through the horrendous wire cuts back to health, then losing him from after effects was a hard ordeal for our family. Peter Paint would accompany Koko to the water hole, supporting him as he slowly and stiffly made his way with his badly swollen legs. The swelling got so bad he began sweating blood. Our trusted vet who had helped us with Koko for months, kindly advised us that the most humane thing we could do was to put him down, which we did.

I thank God for creating the horse, and allowing us to train them and develop special bonds. Working with horses teaches patience, kindness, firmness, respect, and compassion—

and—that renegades exist in the horse world too, that will hurt you no matter how well-meaning you are. This animal has provided more non-food service to mankind than any other. The horse has provided transportation and energy to allow mankind to expand civilizations and conquer new lands. In our own country's history, the horse has profoundly influenced the advancement of cultures. First with the Native American hunting, war, and nomadic culture, and then with the expansion of civilization into the American West.

I marvel at the limitless ways God has used the horse through time and how He's not done yet: *"And the armies which were in heaven followed him upon white horses, clothed in fine linen, white and clean." —Revelation 19:14*. I can't help but smile, thinking how fun it would be to make that trip on a new-spirited Alpo.

CHAPTER 18

Who We Have in Life

"No man will make a great leader who wants to do it all himself or get all the credit for doing it." Andrew Carnegie

I will be the first to admit that nothing I've written about would have happened without God and the people He surrounded me with. Much of this memoir is about values and beliefs, character, integrity, work ethic, and following through on dreams. I owe a debt of gratitude to key influencers in my life and to mentors and friends who stood by me for many years. The bad part about naming names is that I will surely forget someone. For that, I apologize.

"It's not what we have in life—but who we have in life."

I was fortunate to have a profound family influence from my grandfather, Joseph Marxer, who homesteaded on lands near where I grew up, and my Grandma Hazel. As a tiny infant, she traveled with her family in a horse-drawn covered wagon from southern Alberta, Canada, to the prairies of Montana, settling near my family's home place. My grandparents were practically fixtures in our farm home during the summer and fall harvest seasons, helping my parents and establishing a work ethic example for all of us.

My parents, Dale and Shirley Marxer, carried on my grandparents' examples of character, integrity, and responsibility. They loved all five of us kids unconditionally and sacrificed a great deal for us in providing a living and opportunities for personal growth. They showed us and taught us how to have passion and commitment to family, the land, and the community around us.

Of all the speeches I've given across the country, the only one I ever wrote down was my dad's eulogy when he passed away at eighty-two. When researching for Dad's tribute, we discovered that Dad had contributed over 300 years of volunteer service to his fellow man, mostly centered around conservation and care of the land. Mom isn't far behind, with over forty years as a 4-H leader. She also served our community on civil defense boards and other volunteer capacities. Engaging in our local communities and industry-related organizations and commissions is an integral part of securing the future for our families and maintaining a grassroots influence in society.

The circle of community influencers for us siblings included county agents, neighbors, and teachers—especially vocational agriculture (VoAg) teachers, and 4-H leaders. We were all ten-year 4-H members and continued to volunteer service as 4-H leaders or in other community programs in our adult lives.

In 1980 I met the most excellent mentor and influencer of the ages when I accepted Jesus Christ as my Lord and Savior. He's my ever-present friend, guide, counselor, and helper in my life, marriage, and work. The teaching I sit under and the great preachers and evangelists we've been privileged to know have tremendously influenced my life and that of my family.

> *"Happy is the man that findeth wisdom, and the man that getteth understanding." Proverbs 3:13*

Tom Griggs was the number one influence during my first two years with Matador Cattle Company. Tom was the cow foreman who hired me and gave me a chance. He taught me a great deal about handling large herds of cattle in big open country and how to be a cowboy on a big outfit. Being immersed in the lifestyle of working with animals and the range, a cowboy quickly learns total respect for the elements that no man controls.

Tom was a skilled roper on the ranch or at team roping and calf-roping competitions. I was hungry to develop my roping skills, and Tom mentored me, especially in tie-down calf roping, which I enjoyed more than eating. Tom was old-school in many ways. He seldom explained things in detail but would show you by example. Keep your eyes, ears, and mind open and your mouth shut; you could learn a lot. It is said that Marion's brother, Jim Cross, from the Texas Matador, was very similar.

Marion Cross was one of the most influential people during my career. He was an exceptional example in many ways and mentored me when he probably didn't even know he was. There were many times when he purposefully took the time to help me learn and explained things to help me in the future. One of Marion's best attributes was his patience, which likely came from his long experience managing sheep. Most people working with sheep have developed more tolerance than a cowboy, which coincides with how they manage people. Sheep you lead, cows you chase. Marion was not a chaser but a quiet leader.

Marion and I shared several levels of transitions as the company introduced us to new things and tried to push us along into the 1990s. We went from an office with a typewriter, electric adding machine, and a dial telephone to entering the computer and technology age.

Marion committed and followed through when he recruited me for cow foreman, recognizing my concern that my family life and raising our three little kids would suffer. He committed to helping us make it work, and he did. Career-wise, Marion and Jackie experienced many of the same things that Sue and I experienced on the ranch that was probably unique to them and us. Even the end of our careers had similarities.

August L. "Gus" Hormay, a range scientist and the founder and developer of rest-rotation grazing, significantly influenced my range management philosophy. Lessons learned from Gus came from time spent together, his photo plots, and the data he recorded. We were able to observe the positive change in the land over time. The results of that demonstration project influenced me throughout my career in how I approached my land and resource management responsibilities.

Gus loved Sage Creek. The bond he developed with that land over three decades of recording changes on his last surviving major demonstration area, was probably the highlight of his career. Gus requested that his ashes be scattered there upon his death. I, and two others who appreciated his life's contribution as much as I did, were privileged to honor Gus's final request.

From a corporate standpoint, one person stands out as a tremendous example of an influencer and mentor. Sterling Varner had been with Koch for many years and had served in many roles for Mr. Fred Koch. He spent a lot of time at the Beaverhead Ranch when the company first bought it in 1951, drilling wells and making it productive. Sterling made regular trips to Montana throughout his tenure since he loved fishing and visiting the ranch. He also enjoyed spending time with John Erb, our leading cattle trader, who Sterling had known since the 1950s at the ranch.

Sterling, who coined the term "suit," *was* "the suit" in Wichita, who was over the ranches during the early years. Mr. Varner was also the biggest champion of the ranches. I'm sure he still advocated for the ranches even after retiring. Sterling Varner could be credited with much of Koch Industries' success throughout the years. A genuine and personable person, Sterling had a way about him of making even the lowest-level employee feel good and appreciated. It didn't matter if it was the ranch manager or the sheepherder in a sheep camp; Sterling showed a sincere interest in the employee's work. He is one of two people I've admired and attempted to emulate throughout my career.

Sterling made numerous visits to the ranch throughout my tenure, and I enjoyed many conversations with him. Near the end of his career, when visiting the Matador with a group of other folks, we were all discussing the history of Koch's ranches. Sterling was speaking very passionately, explaining to another person in the group what the ranches were like and how they performed. One statement he made always stuck with me. "The ranches get along pretty well as long as the suits in Wichita stay out of the way."

Coming from Mr. Varner, that direct admission was quite profound to me. Sterling Varner had been that "suit" over the ranches for a long time and probably knew the entire ranch history better than anybody in the company. He recognized that the suit could be most effective by being an encourager and enabler—a leader rather than a boss and leading rather than driving.

James Palmer, Ray Marxer, Sterling Varner, Bob Kilmer, John Erb (non-Koch), and Larry Angell at Matador's Beaverhead Ranch cookhouse, Dillon, Montana, in 2001. The five Koch employees represent over 200 years of company service.

In the summer of 2021, Sue and I had the pleasure of visiting the ranch Sterling put together following his career at Koch in the Flint Hills of Kansas. Our son-in-law and daughter, Gregg and Kristy Hoy, help during the shipping of yearling cattle. Sterling took great pride in his ranch, and it shows. Sterling and his wife, Paula, have passed away, but their passion and care for the land continue with their son, Richard, who is now the owner. We had the pleasure of meeting Richard and his wife and enjoyed a stimulating conversation about cattle and reminiscing about his folks. With family ties in Kansas, we plan to be annual visitors.

A distinction immediately apparent to us was how Sterling had structured his own ranch employee management, a structure that successfully continues under Richard's ownership. Sterling Varner worked for a large corporation for over forty years, a corporation that discouraged family involvement and often invoked nepotism policies. He recognized that while there is a place for nepotism policies, a ranch is not the place.

On his own ranch, Sterling discovered the fallacy of forbidding employees' families to be involved. Early on, he found the most successful thing he could do was hire some families to look after his place. Sue and I were pleased and gratified to spend time there and watch about three generations of family working together to gather and ship those yearlings. They were happy, productive, and obviously loved being a part of the ranch—as much, or more, than perhaps even Sterling and Richard.

A fine gentleman by the name of Jack Turner from Sayre, Oklahoma, and Oklahoma City, was a significant help and provider in maintaining the right Hereford influence in our cow herd. The Beaverhead Ranch became acquainted with Jack through Larry Angell and Bob Kilmer, who both went to college in New Mexico. One of their professors was acquainted with Jack Turner since they were all involved with the breed of Hereford cattle.

Jack was probably one of the best cattle breeders I've ever met. Cattle breeding was a science and a real passion of his. Jack Turner understood profit drivers for a cow, and his goal was to produce bulls to provide the perfect balance of traits. Once I began purchasing bulls for the Beaverhead Ranch, Jack became our go-to guy, leading us to partner with him on some Hereford bulls for breeding and AI collection. Jack supplied semen to the Duncan Hereford Ranch in Joplin, Montana, and we were able to get some good Hereford bulls from the Duncan Ranch at their annual sales.

While Jack was an exceptional cattle breeder and businessman, his genuineness was the most influential thing about Jack Turner. Like Sterling Varner, he was genuinely interested in the people he met and with whom he stayed in contact. Jack was a Christian and a very humble man with whom spending time was a joy and time well spent. Besides Sterling Varner, Jack Turner was the other gentleman I tried to emulate throughout my life.

Grant Bowen was a remarkable man I met at Sage Creek early in my career. We were in his sales region, and he would come by and visit, bringing information and different kinds of feed supplements he was selling. Grant was very effective and another of those genuine persons who cared about people. He was similar to Sterling and Jack in the sincere, caring way he showed genuine interest in people and made somebody feel good. Grant was a humble man and like a father figure to me. He had worked for different large corporations marketing feed products throughout his entire career. I trusted him for thoughtful advice, and Grant mentored me in dealing with the challenges of working for a corporate business. He would give me a little foresight of things to expect and examples of how he dealt with them. Grant enjoyed his job and was over seventy-five years old when he retired.

Stan Parsons, who founded and taught the Ranching for Profit School, profoundly influenced my approach to business that would serve the Beaverhead Ranch and the company well. The principles we learned from Stan Parsons and put into practice returned tremendous results for profitability and sustainability across the ranch.

Three individuals who were my contemporaries on the other Koch ranches were a big part of my career and life with the Matador. We understood each other, shared many similar challenges, exchanged ideas and advice, and appreciated each other's friendship. Each of the three served as president of Matador Cattle Company in Wichita at some point in their career. Larry Angell was responsible for much of the growth of Matador Cattle Company and the personal development of the rest of us. Larry served many different roles within the ranches and with other Koch businesses.

Bob Kilmer also served many roles within the ranches and managed the Texas Matador for several years. During the Koch Beef venture, Bob, a kind, reserved, country gentleman and a cowboy rancher, was recruited to be what I've always considered the "poster boy" for the company. Bob was often the face used to portray the Koch image.

James Palmer had similar roles throughout the company and spent time managing the Texas Matador and the Spring Creek Ranch. The four of us were a core group representing the ranches, who grew up together in and with the company. It was a blessing to spend nearly forty years together working with Christian men.

Beaverhead Ranch could not have run so smoothly without our office help managing the paperwork, phones, and bookkeeping. Several secretaries worked through the years to keep our office functioning smoothly. Besides Wanda Everitt, Renae Detton and Janet Brown had stints as secretaries. Deb Harrison worked for us for a long time and was top-notch help.

A few others worked short stints, including Rebecca Eckhardt, a WMC college student.

Montana's Matador Cattle Company could be productive and profitable internally, but it took the support of good neighbors, services, and suppliers to make it all work. Our neighbors were first-rate, and the support help we got from the community provided much needed and appreciated help. Several of the agency people we worked with were solution oriented and helpful in managing intermixed lands, wildlife, and streams to benefit everyone.

Our longtime veterinarian and friend, Tom Williams, provided health care for our livestock. Dr. Bill Hawkins provided health inspections on thousands of cattle for shipping for years, along with Jim Campbell, our brand inspector. Tracy Marquis from Walco was a dependable and helpful provider of animal health products for many years.

Another group of partners was our truckers, who hauled many thousands of cattle. Road and weather conditions were not always ideal and sometimes flat-out challenging, but they always got the job done. Truckers included Fay Riley, Roger Cleverly, Jim Tenney, Leonard Love, Dan Hill, Billy Grosse, and numerous others.

Schuett Farms, who did our custom haying and farming for the Beaverhead Ranch farm, was a dependable service provider from 1995 to the 2020s. Intermountain Irrigation and R. E. Miller and Sons Excavation contributed to the farm's success.

By far, the people most influential in my life, the ones that have made every step of it worthwhile and made it fun to go to work every day, are my wife, Sue, and our children,

Clayton, Kristy, and Anna. They were my support, my inspiration, my joy, my courage, and my determination. They were also a very capable and dependable crew whenever needed. In later years when Todd was herd manager on the Blacktail side, we often helped Todd, as did assistant manager, Kyle Hardin. Most of the cowboy crew spent summers taking care of the Centennial Valley side of the sprawling ranch, and having our own little crew on the Blacktail side of the mountain reduced the need for excessive travel when a large crew wasn't needed.

I don't know where I would have ended up without my family. I had entertained the idea of leaving the ranch twice before we ever left Sage Creek, but they were the best incentive to tough it out.

Not only was my family good help, they loved to have fun—often at my expense. Sue did not open the tight wire gates. She would scoot into my seat to drive the truck and trailer through while I opened, then shut the gate behind her. Without fail, she would wait until I was walking across in front of the pickup and, at the perfect moment, would honk the horn. I could be expecting it, waiting for it even. But I jumped every time to her great amusement.

Marxer family - Anna, Sue Ray, Kristy, Clayton

One spring, when Clayton and Kristy were home on break, Sue and I and the three kids drove to Middle Jake meadows to see how they were looking, and what kind of wildlife was out and about. Chubby, our dog, rode in the truck box.

We were having a fun visit when suddenly I spotted a big bull moose on the edge of the willows, one hundred yards or so below the road. Sue forgot her camera, but I had my digital point-and-shoot on the seat and pulled it out as I came to a stop on the road above the moose. "Shoot!" I grumbled. "I'm too far away."

"Okay, you guys stay here and be quiet," I told the family. "I'm going to see if I can get a little closer." Sue was the one I always worried about getting too close for pictures. I'd been chased by moose twice—once in our own yard—and knew better.

I stealthily crept toward the moose, stepping carefully between swamp bumps, when I noticed Chubby had decided I might need his help and was right at my heels. We got close—almost enough to fill my viewfinder, and I was pretty pleased with myself since Sue always told me I needed to zoom in when I was taking pictures. I took two or three shots to be sure. I'd just turned around and was casually strolling back to the truck, when suddenly the door flew open and Clayton jumped out frantically waving his arms and yelling.

"Dad!" he shouted excitedly. "Look out! The moose!"

I never even looked behind me. With my heart and adrenaline pumping, I sprinted through the swamp bumps like I had wings on my feet. I breathlessly reached the truck to find my entire family practically rolling on the floor with laughter.

I looked back, and the moose had wandered back into the willows. Chubby was standing next to me with his tongue hanging out and grinning at me like even he found it hilarious.

And it was. Hilarious I mean. I've always been able to laugh at myself, and I thank my family for keeping me humble and never wavering in their love and support. By the grace and blessings of God, we had a golden life on that ranch called the Matador Cattle Company. I don't know what happened to those moose photos, but we always enjoy recalling this story, and so many others that made life special and unforgettable.

Kids, never give up on your dreams. The wilder the ride, the sweeter the satisfaction. For myself—I will always be the cowboy God enabled me to be.

Acknowledgements

To our son, Clayton Marxer, AKA The Adventure Cowboy. Thank you for sparking this memoir. Your thoughtful and succinct Facebook post in December of 2021, concerning the sale of the ranch where you were born and raised garnered such tremendous interest and wonderful response that we realized much of the Montana Matador Cattle Company history would die with us. One big difference between corporate and multi-generational family ranches, is that corporate ranch employees come and go and most never know the history. There is no heritage. Clayton, Kristy, and Anna, this is your heritage.

Thank you to the great team at self-publishingschool.com for direction, instruction, guidance, coaching and connections; and to Margaret who deftly got us pointed in the right direction. You've made this published memoir a reality rather than a languishing thought.

To everyone who encouraged me with your interest—especially Debbie at The Bookstore in Dillon—thank you. Debbie's knowledge and enthusiasm to help launch this memoir with our very first book signing was invaluable. Having this book in my hands is a joy beyond anything I ever imagined. I hope and pray it can be a help in some way to each one who reads it.

And, most importantly, to Sue, my lovely wife who is rarely in pictures because she's the one taking them. Thank you for your ever-present camera, and for helping to poke and prod my words and stories into one comprehensive memoir encompassing life on so many levels. I am still amazed, after more than forty years, that you are mine!

Braxton, Ray & Sue, Cody & Clayton, Alexa Marxer, Kristy & Gregg Hoy, Isaac & Anna Ingram w/ Billie Sue. Not born yet: AJ Ingram & Monte Marxer

MORE

Every review is important and helpful. Please take a couple of minutes and leave a review for "Cowboy in a Corporate World" wherever you bought the book. Short and simple is fine. How did you like the book? Something that resonated? Recommend? That little kindness is the greatest thanks you can give any author and is so appreciated.

Visit www.raymarxer.com to see what's new, and sign up for occasional updates, promos, and notices. Signed, jacketed copies will be available only on the website or at book-signings.

CPSIA information can be obtained
at www.ICGtesting.com
Printed in the USA
BVHW040802250922
647569BV00004B/13/J